WILHELM REICH

# The Bion Experiments
## On the Origin of Life

TRANSLATED FROM THE GERMAN
BY DEREK AND INGE JORDAN

*Edited by Mary Higgins
and Chester M. Raphael, M.D.*

FARRAR STRAUS GIROUX
NEW YORK

Copyright © 1979 by Mary Boyd Higgins
as Trustee of the Wilhelm Reich Infant Trust Fund

The original German text, Die Bione, copyright 1938 by
Sexpol-Verlag, renewed 1966 by Mary Boyd Higgins
as Trustee of the Wilhelm Reich Infant Trust Fund

All rights reserved

Printed in the United States of America

Designed by Irving Perkins

First Octagon printing, 1979
First Farrar, Straus and Giroux paperback printing, 1979

Library of Congress Cataloging in Publication Data
Reich, Wilhelm, 1897–1957.
The bion experiments on the origin of life.
Translation of Die Bione zur Entstehung des vegetativen Lebens.
Bibliography: p.
1. Spontaneous generation. 2. Electrophysiology.
3. Orgonomy. I. Higgins, Mary. II. Raphael,
Chester M. III. Title.
QH325.R4413    577    78-27043

*The Bion Experiments*

ALSO BY WILHELM REICH

*The Cancer Biopathy*
*Character Analysis*
*Early Writings, Volume One*
*Ether, God and Devil / Cosmic Superimposition*
*The Function of the Orgasm*
*The Invasion of Compulsory Sex-Morality*
*Listen, Little Man!*
*The Mass Psychology of Fascism*
*The Murder of Christ*
*People in Trouble*
*Reich Speaks of Freud*
*Selected Writings*
*The Sexual Revolution*

*The Bion Experiments*

*Love, work, and knowledge are the wellsprings of our life. They should also govern it.*

**WILHELM REICH**

# Contents

Preface     3

### PART ONE    *The Experiment*

1. *The Tension-Charge Formula*     19
2. *Bions as the Preliminary Stages of Life*     25
   - VESICLE FORMATION IN SWELLING BLADES OF GRASS     26
   - THE TRANSFORMATION OF GRASS AND MOSS TISSUE INTO FORMS OF ANIMAL LIFE     31
   - CONFIRMATION OF THE VESICULAR CHARACTER OF THE FLOWING AMOEBAE     38
   - MOTILE VESICULAR EARTH CRYSTALS AND EARTH BIONS     39
   - EGG-WHITE PREPARATIONS (PREPARATION 6)     54
3. *The Culturability of the Bions (Preparation 6)*     64
   - ELECTRICAL EXPERIMENTS     74
4. *The Beginning of Control Experiments by Professor Roger du Teil at Centre Universitaire Méditérranéen de Nice*     84
5. *Culturability Experiments Using Earth, Coal, and Soot*     99
   - ELIMINATION OF THE OBJECTION THAT PREEXISTENT SPORES ARE PRESENT     99
   - THE INCANDESCENT COAL EXPERIMENT     107
   - CULTURES OF SOOT HEATED TO INCANDESCENCE. THE BIOLOGICAL INTERPRETATION OF BROWNIAN MOVEMENT     110
6. *Control Tests and Instructions for Verifying the Bion Experiments (Summary)*     115
   - POSSIBLE FUTURE STUDIES     118

PART TWO  *The Dialectical-Materialistic Interpretation*

7. *The Problem of the Mechano-electrical Leap*    125
    CHEMICAL PRECONDITIONS OF THE TENSION-CHARGE PROCESS    125
    ELECTRICAL CHARGE AS A CHARACTERISTIC OF COLLOIDS    127
    ELECTRICAL CHARGE AS A PREREQUISITE FOR VESICULAR MOVEMENT    129
    INDIVIDUAL FUNCTIONS AND INTEGRAL FUNCTION    135

8. *An Error in the Discussion of "Spontaneous Generation"*    140
    SUMMARY    145

9. *The Dialectical-Materialistic Method of Thinking and Investigation*    146
    THE BASIC METHODOLOGICAL APPROACH TO OUR EXPERIMENTAL WORK    146
    THE DIALECTICAL-MATERIALISTIC LAW OF DEVELOPMENT    153
    SOME REMARKS ON BIOGENESIS    171

*Appendix*    178
    PRODUCTION OF BIONS FROM STERILIZED BLOOD CHARCOAL    178
    THREE SERIES OF EXPERIMENTS BASED ON THE TENSION-CHARGE PRINCIPLE, BY ROGER DU TEIL    179

*Bibliography*    193

# Editors' Note

*The Bion Experiments* is Wilhelm Reich's detailed account of his laboratory investigations on the origin of life. Published originally in Oslo in 1938, it set off shock waves of indignation and controversy culminating in a savage newspaper attack that made Reich's continued sojourn in Norway impossible and caused the dismissal of his collaborator, Professor Roger du Teil, from his university position in France.

The protest that greeted these experimental findings and the report itself were soon drowned out by the Second World War, and although Reich summarized the results of his experiments in a later work, *The Cancer Biopathy*, full information has not been available.

In preparing this new edition, we have been torn between our wish to carry out instructions written by Reich in 1947, from the vantage point of his knowledge ten years after the completion of this report, and our concern for historical accuracy. He wrote: "The philosophical article of du Teil's should be left out of the reprint of *Die Bione*. It will be necessary to add a preface to [the] publication of *Die Bione* which would point out the progress made in orgone research since its first publication, and, most important of all, to change the term 'dialectic materialism' to 'energetic functionalism' which it truly and really is. I could no longer afford, as I did ten years ago, to have my method of thinking and research termed dialectic materialism, since (1) the socialist and communist parties are still using the term without giving it any meaning; (2) I don't wish any more to be confused with the Marxist political parties; and (3) energetic functionalism of today has as much to do with dialectic materialism as a modern electronic radar device with the electric gas tube of 1905."

The article by du Teil has been omitted, but we found it impossible to change the term "dialectic materialism" without creating serious confusion and damaging the integrity of the work. It therefore remains as Reich used it at this period of his scientific development. The reader should be reminded, however, that Reich's continued exploration of the path opened up by these early bion experiments led to his discovery of biological energy, the orgone, and enabled him to develop his formulation of dialectical materialism derived from Engels into energetic, i.e., orgonomic, functionalism.

                                Mary Higgins
                                Chester M. Raphael, M.D.

**Forest Hills, N.Y., 1978**

# Preface

It is with some trepidation that I make known these experimental findings on the origin of vegetative life. It is not that I am worried about the correctness or accuracy of the data given, even though here or there an insignificant error or an awkward phrase may have crept in. All the findings described in this comprehensive, yet not definitive, report were confirmed hundreds of times. I have omitted any observations that were not verified and I have gone to great lengths to describe the method as precisely as possible, so that it can be tested by others. If the instructions are followed more or less correctly, it is impossible to miss the basic phenomena such as the vesicular disintegration of matter upon swelling or the culturability of the bions. I fully realize that the same findings are open to other interpretations than my own. For this reason I have carefully separated the factual report in Part One from the interpretation in Part Two.

I am concerned that I might be criticized as immodest for drawing the conclusions that I do from these experiments. I stayed within the bounds laid down by eighteen years of clinical work on the functionally diseased organism and ten years of intensive study of the relevant biological and physiological literature. The sections on colloids and on the dialectical-materialistic method of research were finished many years ago but lay unpublished in my desk drawer. They represented attempts to link my practical experience as a psychotherapist with my general biological studies. I had become directly aware of the connection with psychoanalytic knowledge, on the basis of my orgasm theory, when in 1926 I was asked to review a book by Fr. Kraus on the pathology of personality (*Syzygiologie*) for a scientific journal.

I did not suspect that ten years later I would be given the opportunity to verify natural philosophical assumptions and the dialectical-materialistic method in such a way, although I knew, of course, that the orgasm theory touched on the "life problem." What I submit here is not a random discovery, but a development over a period of years of work on the problem of the autonomic function. Step by step the fundamentals of a theory of biogenesis, which had to be worked out in full, were revealed. I have to admit that the facts I discovered seemed incredible at first. But fact after fact came to light and each one confirmed the picture that I had already formed from clinical studies of the life function and its disturbances. By the time I published "Experimentelle Ergebnisse über die elektrische Funktion von Sexualität und Angst" in 1937, the results of the bion culture experiments were already available. Now that I have decided to publish them, I have at my disposal additional data in a related area which confirm and represent a continuation of these experiments.

The techniques which I used in the experiments do not differ from those customarily employed for bacteriological sterilization. However, the arrangement of the experiments as well as the methods of interpretation and the conclusions that are drawn differ considerably from the norm. The experiments were all based on the fundamental formula which I had discovered in the course of my research in the field of sexuality. The analytic method follows the laws of dialectical materialism. Marx had added the element of materialism to the Hegelian dialectic, but the method was first used in a natural scientific context by Engels; it then found a new application in psychology and the process of sexuality. The principles of the method became more refined and new ways of obtaining knowledge were revealed as, for example, in the "dialectical-materialistic law of development." From Freud I adopted the hypothetical equation of life impulses and sexual impulses. Once I had succeeded in refuting his theory of the death instinct and in developing my orgasm theory, I was able to proceed to experimental biology. The experimental proof of the identity of the sexual energy process and the life energy process is thus simultaneously a confirmation of Freud's hypothesis.

At this point I would like to express my warmest thanks to Professor Roger du Teil for the incomparable friendship he has given me throughout our collaboration. Whatever effect his efforts to draw the attention of biologists and bacteriologists to this work may have, his active participation in the experiments has become an organic part of the entire series of studies. This is clear from the text that follows.

I am also aware that the experimental solution of the question of spontaneous generation satisfies many needs throughout the scientific world. Similarly, I know that I will have to face some sharp opposition. However, the back and forth of argument and counter-argument constitutes the very essence of scientific work. What is more, every objection leads to progress if the fundamental problem is correctly grasped.

My work "Der dialektische Materialismus in der Lebensforschung" (*Zeitschr. f. pol. Psych. u. Sexök.*, No. 3, Vol. IV, 1937) gives a historical analysis of the development of the problem. It also points out the connections that exist between this problem and sociological questions. I have left for future publication the details of many studies and also the analysis of related questions.

I am particularly grateful to Professor Harald Schjeldrup for having made possible and actively assisted in carrying out the initial physiological electrical experiments at the psychological institute of his university. Without his assistance, even in general matters, I would have had to overcome many more problems.

Extraordinary material difficulties were encountered in setting up the laboratory operations. The Rockefeller Foundation in Paris refused its support. It would not have been possible to conduct the experiments at an official establishment engaged in other work, and I would never have been able to manage alone. Therefore, I should like to take this opportunity to thank publicly all those who made the undertaking possible in the face of difficult odds. Above all, my thanks are due to my friend Sigurd Hoel, whose advice often kept me from losing faith in my ability to see the project through. I am also grateful to our friend Dr. Odd Havrevold, who set up the laboratory in which the experiments were conducted, provided general practical assistance, and solicited contributions. In addition, my thanks go to those who helped me carry out the bacteriological, cinephotomicrographic, and physical-chemical work and who, through their initiative and drive, helped me overcome many obstacles. Much more would have gone wrong without the active material support given the institute by my colleagues in the field of character analysis; they helped me to set up and maintain the entire operation: Dr. Lotte Liebeck, Dr. Nic. Hoel, Dr. Ola Raknes, Dr. Tage Philipson, Dr. Leunbach, Ellen Siersted.

However, these specialists were not able to provide large sums of money and their efforts alone would not have been sufficient. (The equipment for the biological laboratory alone cost approximately 60,000 Norwegian kroner. At the present time it costs approximately 2,000 Norwegian kroner per month to operate the laboratory.) My work was decisively aided by large contributions from Mr. Lars Christensen (Oslo), Mr. Rolf Stenersen (Oslo), and Constance Tracey (London).

The overall project was greatly assisted by the administrative staff and in particular by my secretary Gertrud Brandt, who tirelessly and efficiently maintained order in my wide range of activities. The head of our publishing house, Mr. Harry Pröll, supervised the production of the book with great care and diligence.

The Institute was founded by Norwegians. The extraordinary hospitality of the Norwegian people has provided a fertile background and basis for my work, full responsibility for which is mine; Norway is a country that has been able, by and large, to keep the emotional malaise of the world at bay.

## THE ESSENTIAL LABORATORY EQUIPMENT
(Figures 1-11)

The complicated experiments designed to determine the microbiological and electrical properties of the substances, as well as of the various types of bions, required equipment which was adapted to specific purposes or, in some cases, which had to be specially created.

*The microscope*

At present our institute possesses three large Reichert "Z" microscopes and one Leitz research microscope. With the Reichert microscopes it is easy to achieve a magnification of up to $3750\times$, as a result of the inclined binocular tubes, which increase the normal magnification by 50 percent. When a special Leitz $150\times$ apochromat lens is used in conjunction with a $25\times$ compensating ocular and the inclined binocular tubes, it is possible to achieve a magnification of up to $4500\times$, but with great difficulty. Dark field examinations were carried out at approximately $300\times$ to check for motion and at $1200\times$ to assess the coarse structure and the type of motion. Furthermore, observations were conducted at approximately $3000\times$ to determine the fine structure of the organisms and the vibrations inside their body mass visible only at this magnification. In order to assess the internal movements reliably, a dark field condenser, manufactured by Reichert of Vienna, was also used. With this device it is possible to make observations in a dark field at approximately $3000\times$.

This manipulation is very complicated and requires lengthy preparations. Many characteristic processes could be seen only by using the Reichert "Z" microscope. This microscope revealed phenomena which would certainly not have been visible using a straight single-tube instrument or even one with non-inclined binocular tubes. *It is not really possible to verify the findings unless the same optics are used.*

## Cinephotomicrographic apparatus

Each new process that was observed, if it proved to be typical, was immediately filmed. Two types of cameras were used. We had a CK Pan Film Camera, Kodak (F I, 9), which permitted a speed of eight frames per second; i.e., the motion was accelerated to twice normal speed. On the average, filming was done between magnifications of 300× and 1500×, using a single-tube microscope and fixing the camera lens directly above the ocular of the microscope. By means of a special device it was also possible to film structures which moved only slightly; in this case, a microscope with inclined binocular tubes was used at 2300× and the camera was mounted on one of the oculars.

The large Cine Kodak Special Camera (F I, 9), used for time-lapse photography of developmental processes, permits single exposures to be made; also, the light intensity and exposure speed can be adjusted very accurately.

Two time-lapse devices were used. One was an electric release control for the Cine Kodak Special manufactured by the Eastman Kodak Company. By switching various relays, one could accelerate the motion in the following order of magnitude:

```
 4× normal speed   (4 frames per second)
 8×    ”      ”    (2    ”      ”    ”  )
16×    ”      ”    (1 frame     ”    ”  )
32×    ”      ”    (1    ”   every 2 seconds)
48×    ”      ”    (1    ”      ”   3   ”  )
64×    ”      ”    (1    ”      ”   4   ”  )
80×    ”      ”    (1    ”      ”   5   ”  )
96×    ”      ”    (1    ”      ”   6   ”  )
```

In order to speed up the motion ninety-six times, one meter of film was exposed in thirteen minutes and twelve seconds. This apparatus was used for filming developmental processes and forms of motion which could still be seen at high magnification, although with some effort. For filming processes of develop-

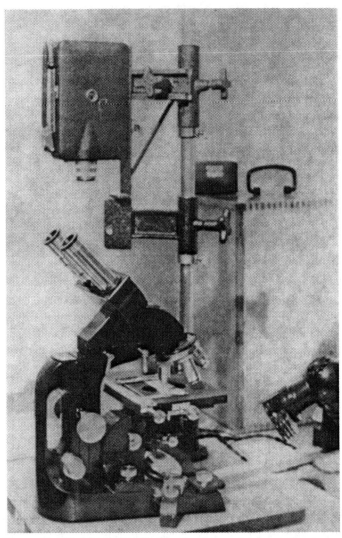

Figure 1. The large Reichert microscope for magnifications up to 4500×

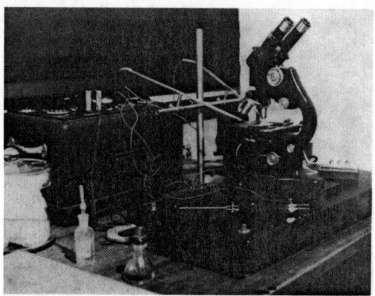

Figure 2. Apparatus for micro-electrical studies

Figure 3. Cinephotomicrographic apparatus (for short and long-interval time-lapse filming)

Figure 4. The two relay control units for the time-lapse filming

Figure 5. The cinephotomicrographic apparatus with motor for time-lapse filming

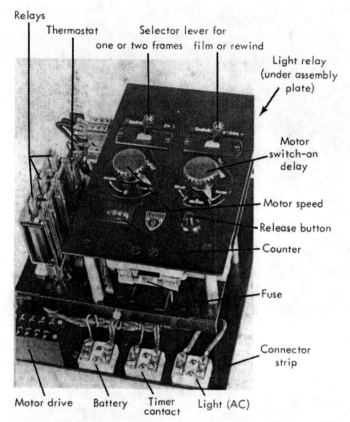

Figure 6. Relay control unit for long-interval time-lapse filming

Figure 7. Contact timer for setting time intervals

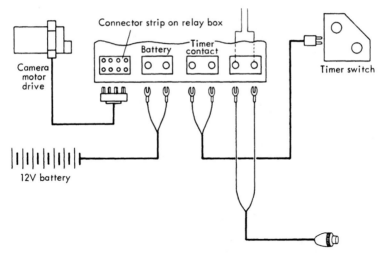

Figure 8. Switching circuit for the long-interval time-lapse apparatus

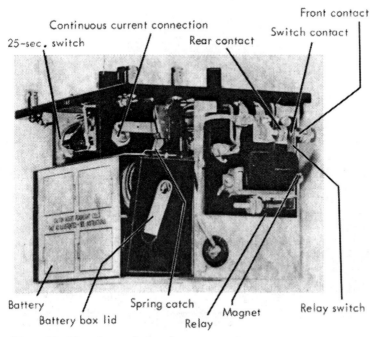

Figure 9. Short-interval time-lapse apparatus

Figure 10. Three-tube amplifier and silver electrodes

Figure 11. Oscilloscope, roll film apparatus, and non-polarizable screened electrode

ment and movement not directly observable, a time-lapse device manufactured by Askania (Berlin) was used. The system of switches and relays permitted the following speeds to be selected:

| | | | | | | | |
|---|---|---|---|---|---|---|---|
| One frame every | | 15 seconds | | (240× | normal | speed | ) |
| ” | ” | ” | 20 ” | (320× | ” | ” | ) |
| ” | ” | ” | 30 ” | (480× | ” | ” | ) |
| ” | ” | ” | 40 ” | (640× | ” | ” | ) |
| ” | ” | ” | minute | (960× | ” | ” | ) |
| ” | ” | ” | 5 minutes | (4800× | ” | ” | ) |
| ” | ” | ” | 10 ” | (9600× | ” | ” | ) |
| ” | ” | ” | 15 ” | (14400× | ” | ” | ) |
| ” | ” | ” | 20 ” | (19200× | ” | ” | ) |
| ” | ” | ” | 30 ” | (28800× | ” | ” | ) |
| ” | ” | ” | 40 ” | (38400× | ” | ” | ) |
| ” | ” | ” | hour | (57600× | ” | ” | ) |
| ” | ” | ” | 2 hours | (115200× | ” | ” | ) |
| ” | ” | ” | 5 ” | (288000× | ” | ” | ) |
| ” | ” | ” | 10 ” | (576000× | ” | ” | ) |

In the last time adjustment in the above table, one meter of film was exposed in fifty-five days and nights. Time-lapse exposures were made at magnifications between about 300 and 1200×.

A specially constructed apparatus (see Fig. 2) was used for the micro-electrical studies. A solid round rod was mounted vertically on a sturdy base; a transverse rod was attached to this vertical rod in such a way that it could be moved. To this transverse rod two glass tubes were attached which could be moved in two directions and through which ran a copper wire. At one end a fine thin platinum wire protruded. The platinum wires were attached to eyes fixed on opposite sides of a trough-shaped container on a slide. This apparatus was connected with a pantostat manufactured by Siemens (Berlin) permitting exact measurements and metering of current down to 0.2 mA.

Later on, all the films were made with the aid of an optical adapter which permitted observations while the film was being exposed. The camera can be mounted over the ocular vertically as well as horizontally. By the summer of 1937 we had prepared one complete film of preparation 8 (development of protozoa) and one complete film of preparation 6 (bion experiment); and one film was near completion: preparations 1, 2, and 3 (preliminary stages of life represented by swelling earth, coal, and soot). The laboratory also possessed all the equipment needed to develop the film.

The electrical potential was measured by an oscilloscope which was connected to a three-tube direct-current amplifier. This apparatus was manufactured by the University Instrument factory in Lund (Figs. 10 and 11).

A complete laboratory with autoclaves (sterilization at 120°C) and dry sterilizer (sterilization up to 190°C) was set up for the bacteriological investigations.

W.R.

October 1937

PART ONE

*The Experiment*

# 1

## *The Tension-Charge Formula*

In this work I will describe my observations made during experiments in which inanimate matter was transformed into bacterial organisms. Let me begin by briefly outlining the theoretical basis for the experiments.

In the course of about fifteen years of clinical work, I came to recognize a formula for the function of the orgasm which was verified in subsequent experiments.* In vegetative life there is a process through which mechanical filling, or *tension*, leads to a build-up of *electrical charge;* this is followed by *electrical discharge*, which, in turn, culminates in *mechanical relaxation*. This phenomenon raised two questions:

1. Does this formula apply only to the function of the orgasm, or is it valid for all vegetative functions?
2. Since the orgasm is an elementary phenomenon of life, the formula expressing it should also be demonstrable in the most primitive biological functions; for instance, the vital functions of protozoa. The basic assumption, therefore, was that the orgasm formula is identical with the life formula. Initially, I was not very optimistic about finding proof of this assumption within a short time. It was quite fortuitous that I was able to solve the major part of the problem relatively quickly and with certainty.

In addition, my clinical and experimental experience had

---

* "Experimentelle Ergebnisse über die elektrische Funktion von Sexualität und Angst" (1937); "Der Urgegensatz des vegetativen Lebens" (1934); both published by the Sexpol-Verlag, Oslo.

raised a series of questions which guided the biological investigations. In the electrical experiments on sexual zones it had been discovered that the vegetative excitations are functionally identical with corresponding directions of flow of electrical current. The vegetative excitations proved to be functionally identical with primary vegetative movements which could basically be divided into two groups: the sensation of reaching out and well-being—i.e., *expansion*—corresponds to actual stretching, as illustrated by erection of the penis. On the other hand, anxiety and a feeling of unpleasure are identical with "retreat into the self"; i.e., with a *contraction* of the biological organism. In marine mollusks which I observed, the alternation between expansion and contraction was startlingly clear. The discharge of electrical energy during contractions of electric fish confirmed my assumption that sudden contraction is functionally identical with electrical discharge. Hence, I felt I could allow myself the mental leap in concluding that electrical charge at the periphery is functionally identical with expansion and a feeling of well-being, while electrical discharge at the periphery is identical with contraction and fright or anxiety. In expansion, to pursue my theory further, the distance between particles is increased by the process of swelling, a process which must be closely connected with the increase in electrical potential. In contraction, the distance between particles decreases as a result of shrinkage; thus, the tissues are more resistant and there is a drop in electrical potential; i.e., a discharge occurs. Logically, it should be possible to experience the physical electrical potential directly, in the form of a vegetative sensation of excitation.

Furthermore, about three years ago, in the course of my clinical work on muscularly hypertonic neurotics, I had discovered the *orgasm reflex*. After the hypertonicity had been eliminated, isolated vegetative contractions in various parts of the body combined to give a single total body reflex which I called orgasm reflex. This is the same phenomenon as the automatic vegetative convulsion that takes place at the climax of sexual gratification. I could only conclude that the autonomic nervous system expands and stretches when pleasure is experienced and contracts in the case of fright. The unity of function of the total organism seemed decisive to me here; i.e., the amoeba lives on in the metazoan in the form of the contractile and expansile vegetative apparatus.

According to this view, the nerves of the organism no longer seemed to be the generators of the impulses, but instead were merely organized transmission paths for the vegetative impulses of the entire body. In the literature I found abundant evidence for the view that the ganglia of the vegetative nervous system function as storage batteries and that the muscles act as discharge apparatuses which produce motion. The body fluid, which in the case of human beings accounts for about 80 percent of the total body weight, must be regarded as the most important medium for the propagation of electrical excitations.

The basic functions of living creatures—namely, expansion and contraction—dominate all life, but they themselves are composed of a complicated combination of individual physical functions. I will go into detail later about the facts revealed by colloid chemistry. At this point I wish to restrict myself to a brief description of an overall system of uniformity, not only within the realm of organic life, but also between organic and inorganic functions. As I have already stated, these were just conjectures which arose out of a large series of clinical and experimental studies.

The biological direction "toward the world" represented in expansion, and the opposite direction "away from the world," "retreat into the self," represented in contraction, seemed to me to have a primitive model in the mechanical act of expansion of a pig's bladder. If a pig's bladder is filled with air it stretches mechanically. The surface becomes tense and strives to return to its original state; the process is similar to that of a taut spring. The internal pressure exerted by the air prevents the restoration of the original state. There are now three possibilities:

> The internal pressure is *less* than the surface tension, so the bladder can be pumped up still further without bursting.

The internal pressure is *equal* to the surface tension, so the bladder assumes a stable spherical shape.

The internal pressure can finally *exceed* the surface tension, so that the bladder bursts.

In the living realm an increase in internal pressure leads to a contraction, as in the urinary bladder, or to constriction and division, as in a cell.

In electricity, I was struck by the antithesis of charge and discharge. In the inorganic sphere mechanical tension and relaxation and electrical charge and discharge are separate functions. The organic or living sphere, however, is governed by a specific combination of the two physical functions: tension → charge → discharge → relaxation. This is the formula for biological functioning.

In chemistry there are certain substances which have a swelling (i.e., tensioning) and a shrinking (i.e., relaxing) effect. When potassium chloride and lecithin act on the tissue, the surface tension increases as a result of the swelling, i.e., expanding effect. When calcium and cholesterin act on the tissue, the surface tension is reduced as a result of the shrinking, i.e., contracting effect.

For the understanding of organic functioning, it is obviously significant that tension and relaxation, swelling and shrinking, stretching and drawing together, charge and discharge, etc., are all combined together in *one* system in the functions of the parasympathetic and sympathetic. In a special study entitled *The Basic Antithesis of Vegetative Life* (*Der Urgegensatz des vegetativen Lebens*, 1934), I described this situation, drawing on the experimental results achieved by other authors. Potassium has the same effect as lecithin, lecithin as the vagus (parasympathetic), and the vagus, finally, as pleasurable excitation, swelling, turgor, increased surface tension, and, as was recently shown, electrical charge. In contrast, calcium, cholesterin, the sympathetic nervous system and unpleasure or anxiety form a functional unit characterized by shrinking, contraction, discharge, and reduction in surface tension.

The following is a comparative table which I prepared four years ago:

| Vegetative Group | General Effect on Tissues | Central Effect | Peripheral Effect |
|---|---|---|---|
| *Sympathetic* Calcium (group) Adrenalin Cholesterin H-ions | Reduction of surface tension Dehydration Striated musculature: flaccid or spastic Reduction of electrical excitability Increase of oxygen consumption Increase of blood pressure | Systolic Heart musculature is stimulated | Vasoconstriction Intestinal peristalsis decreased |
| *Para-sympathetic* Potassium (group) Cholin Lecithin OH-ions | Increase of surface tension Hydration Muscles: increased tonicity Increase of electrical excitability Decrease of oxygen consumption Decrease of blood pressure | Diastolic Heart musculature relaxed | Vasodilatation Intestinal peristalsis increased |

That which biologists and, in particular, metaphysical biologists have so far referred to as "organizing intention," "entelechy," etc., seemed to me to be contained in the jump from individual physical functions to a *combination* of these functions which governs the process of life. Thus, it was possible to replace metaphysical biologistic interpretations by the dialectical-materialistic formulation of life processes (see Chapter 8, Part Two).

The uniformity of organic functioning is regulated by the tension-charge process in both the individual organs and the total organism. The uniformity between inorganic and organic processes is contained in the functions of expansion-contraction and charge-discharge. *The difference between organic and inorganic arises from the specific combination of functions in the organic which otherwise occur singly in inorganic substances.*

From these premises, I proceeded to carry out the biological experiments described in the following chapters.

2

## Bions as the Preliminary Stages of Life

The vegetative currents which I had encountered in the course of my character-analytic work and in my electrical experiments on sexuality struck me as so important that I decided to study them microscopically in protozoa. The Botanical Institute in Oslo supplied me with a preparation of protozoa for this purpose. Since I wanted to prepare such samples myself, I inquired how cultures of this sort could be grown; I did not wish to rely on my own knowledge of biological procedures acquired about twenty years earlier and since largely forgotten. Although I was familiar with Leeuwenhoek's infusoria, I was very surprised to hear that it took only an infusion of water and hay—i.e., semi-dry grass—to produce them. I also discovered that amoebae are often found on leaves that have lain for a long time in stagnant pools of water. I was ashamed of the gaps in my knowledge of biology when I naïvely asked how the animals entered the infusion in the first place and I received the astonished reply that there are "germs of life" everywhere from which the protozoa develop. Obviously, I had deliberately, though unconsciously, "forgotten" the "germ theory." I wanted to make a special study of amoeba cultures in order to familiarize myself with the plasmatic currents or streamings described by Hartmann and Rhumbler which were so important for my theory of vegetative functioning. However, the infusion that I had been given contained very few amoebae; it was, in fact, not so easy to prepare fresh cultures of amoebae. The laboratory assistant at the Institute advised me in my predicament to make my own infusions of hay and to examine them "*after about ten to fourteen days.*" I would then certainly find some amoebae.

My own knowledge of protozoology was woefully inadequate. Nevertheless, trusting my basic theoretical knowledge of biology and relying on the experience that I had gained in the last few years in the therapeutic and experimental investigation of the orgasm function, I took the risk of venturing into what was for me new territory. To begin with, I deliberately avoided making a new study of specialized biological literature, so that I could pursue my observations in an unprejudiced manner. I had one of my assistants compile a review of the literature.

## VESICLE FORMATION IN SWELLING BLADES OF GRASS

Paramecia, various types of amoebae, and, among other things, wormlike wriggling objects were easy to observe in the preparation. I was immediately struck by the vegetative plasmatic current and it was easy to identify both the leisurely serpentine movements, progressing in slow motion, and the plasmatic streaming as the tension-charge processes that I was looking for. As an experiment, I passed a current of about 0.5 milliamperes through the preparation and saw, as Rhumbler and Hartmann had described, that the plasmatic streaming accelerated when a weak current was applied. The amoebae moved faster and the movement of the granular endoplasm within the cell, when observed at 1500×, became extremely lively. On the other hand, the paramecia seemed to become highly disorganized. Their rapid and constant locomotion ceased and they started to go around in circles; it appeared as if each application of current had a sudden "traumatic" effect. After the current had been applied uninterruptedly for about three minutes, all motion ceased, with the exception of the amoebae. When I increased the current to about 1.5 mA, the amoebae rolled into balls and also stopped moving.

These observations seemed important to me, but I was still unable to establish a connection between my hypothesis and the change in plasmatic streaming brought about by the application of current. However, there was another way to go about checking the tension-charge formula.

It was clear that living organic tissue took in fluid and, as it swelled, the membranes were mechanically stretched. The so-called turgor of the tissues—namely, the overfilling of blood vessels or hyperemia—consists in principle of nothing more than increased fluid intake (blood, lymph, water). I therefore concentrated my studies on the *changes taking place along the margins of the plant fibers.*

The luminescent green fibers lost their color in the course of about two days and the liquid contained a large number of green chloroplasts. But as the plant fibers lost their color and underwent decomposition, strange changes took place. The previously striated structure of the cells still intact gave way to a vesicular configuration. At the same time, hemispherical bulges or sometimes jagged strips formed at certain points along the edges of the fibers.

The fibers had obviously taken up water and undergone swelling and the chloroplasts had detached themselves; also, the cellular structure of the tissue had started to break down into vesicles. The vesicular structure varied. In some places it was irregular, without a clear border, and in others, a completely regular structure with a border which stood out clearly as the micrometer screw on the microscope was turned. Where there were sharply defined borders, the object looked as if it had been "shaped." The more encompassing the border, the more tautly filled the structure seemed. The vesicles gradually detached themselves from the fibers and floated around in the previously clear liquid. The plant fibers then looked like branches of trees stripped of their leaves (Figs. 12, 13, and 14). The liquid was filled with motionless clumps of vesicles without any defined borders. Their structure was identical to that of vesicles into which the margins of the plant fibers had disintegrated.

I observed the change taking place in both the vesicular structure and the formation of a border in *one* object over a period of four hours. An irregularly structured, boundaryless, vesicular object had formed at the margin of a piece of plant. The object gradually swelled and detached itself from the section of plant. Double refracting margins appeared on the edges. The vesicular structure became more regular and homogeneous and the vesicles refracted light with greater contrast. In its structure the object was almost indistinguishable from a passing amoeba. It assumed a long oval shape and became increasingly taut as the margin became more complete and distinct. Within the object, the vesicles were motionless, but the whole thing had taken on a "shape." The object disappeared when I added water to the preparation.

An amoeba slowly moving by was caught up near the edge of the field, where it then dried up and assumed a spherical shape, at which point it was indistinguishable from the many clumps of vesicles surrounding it.

A further observation pointed in the same direction. Formations which exhibit both the vesicular structure described above and the well-defined margin are seen adhering to the fiber by means of a stalk. Occasionally, such an object can be observed making violent attempts to break away. It jerks away from the substrate but is pulled back again by the threadlike stalk. It is important to note here that only its motility distinguishes it from the lifeless forms, while the shape and structure seem to be the same. The large numbers of vesiculate forms which move around rapidly, as if floating, in the observer's field of vision appear to be of the same type.

Let me summarize my observations:

1. Swelling plant fibers disintegrate into vesicles.
2. Dried-up amoebae have a vesicular structure like the clusters of vesicles.
3. Vesicular formations with definite borders detach themselves from the disintegrating plant.

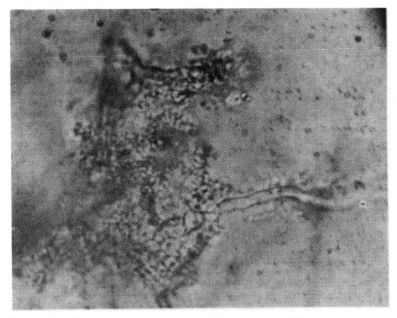

Figure 12. Vesicularly disintegrated blades of grass. Film preparation No. 8, magnification 700×

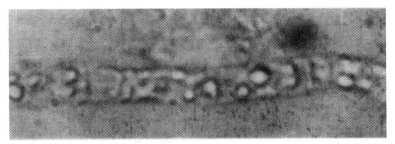

Figure 13. Vesicles inside a blade of grass, 1500×

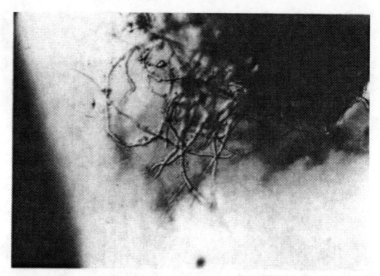

Figure 14. Strands of grass after detachment of the vesicles. Film preparation No. 8, magnification 700×

Figure 15. Sharply defined border of blade of grass from film preparation No. 8, semi-dark field, 300×

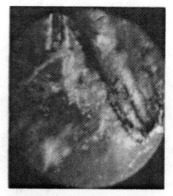

Figure 16. Vesicularly disintegrated grass, dark field, 1000×

*Bions as the Preliminary Stages of Life* 31

4. The distinct border that forms gives the clump of vesicles a definite shape and fullness.
5. Some rapidly moving cells have a sharply defined border and a uniform vesicular structure.

I was forced to assume, no matter how surprising this was, that *the living vesicular ("honeycomb") plasma of the amoeba must be very closely related to the vesicular structure of the disintegrated plants. Could it be possible that an amoeba or other protozoan with a similar vesicular structure is nothing more than a cluster of vesicles enclosed and shaped by a membrane?*

I think it best to let my further observations and experiments speak for themselves.

## THE TRANSFORMATION OF GRASS AND MOSS TISSUE INTO FORMS OF ANIMAL LIFE

When a certain kind of grass is immersed in water, a type of protist forms from the decomposing plant fibers after passing through a series of *preliminary stages*. Immediately after immersion, the blade of grass exhibits a sharply defined margin with individual claw-like hooks (Fig. 15). Over a period of twenty-four to forty-eight hours, this breaks down into vesicles in the manner described (Fig. 16). On the third day I was able to observe spherical to oval-shaped saccules at irregular intervals along the margin of the blade of grass. These forms contained granules and vesicles of various size (Figs. 17, 18, and 19). The small vesicles inside each large membraneously enclosed saccule were indistinguishable from the vesicular inclusions at the edge of the plant. The large round objects were attached to the edge by a stalk. Some of them retained this shape unchanged; some, however, behaved rather strangely. The spherical cluster facing away from the edge of the plant gradually became elongated and, in the process, an opening surrounded with very fine cilia formed at the front. The structure remained stretched for one to three seconds, then it *contracted* and suddenly reassumed a spherical shape. These objects look like round apples or elongated olives, which move while hanging on a branch. Simultaneously, or a little later, they swam around freely in the liquid, sometimes still connected by a stalk to detached fragments of plant fiber. The motion described is the same. The elongation followed by contraction to the spherical shape continues for hours. Occasionally, an object of this sort is seen to move away from the piece of plant during elongation and to stretch the curved stalk. While contracting, some of these formations tend to spring back against the plant. Is this formation already an "animal," or is it still just a "piece of plant"? The question is incorrectly phrased. The object is a *transitional form* because two days later the same preparation contained the same creatures, *detached from the margin of the plant* and swimming freely in the liquid. But this time some of them looked different; they were in every conceivable stage of development. There were still some vesicular spherical structures attached to a plant fiber and stretching and contracting. Some others in the same stage had already detached themselves from the plant substrate. A third group was particularly striking: Instead of contracting back to the spherical shape, they retained their *elongated form;* the "mouth" was wide open and they resembled paramecia. In fact, they moved like the latter, swimming rapidly in all directions. They contained a pulsating vacuole at the front end and vesicles moving tremulously in circles. Those which still continued to revert to the spherical shape made particularly obvious eating motions each time they assumed the elongated form. Once the object had become completely stretched, the mouth was wide open and the individual vesicles in the fluid were seen to move rapidly toward the mouth opening and then through and into the inside of the object. After two or three seconds the mouth closed, the organism contracted back to the spherical shape and the vesicles inside it started to move around vigorously; but I was not able to observe this motion in detail. The elongated object did not swim towards the vesicles in the liquid, but instead it remained stationary and

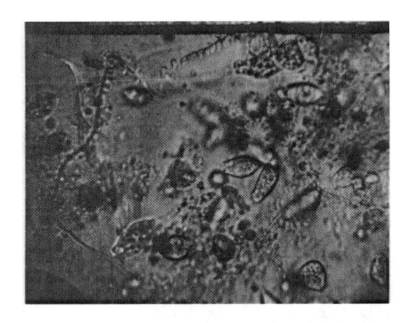

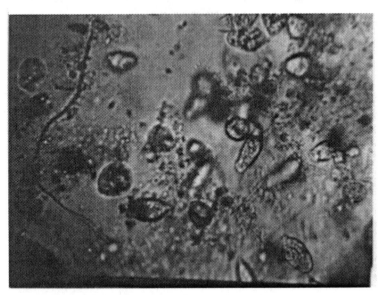

Figures 20 and 21. Org-animalcules on blade of grass. Two stages of formation

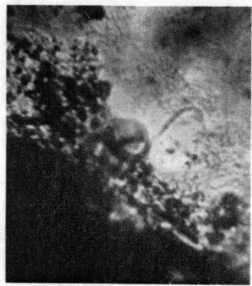

Figure 17. Vesicular marginal structures. Film preparation No. 8, magnification 1200×

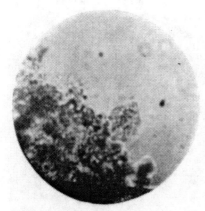

Figure 18. Organized marginal structure. Film preparation No. 8, magnification 700×

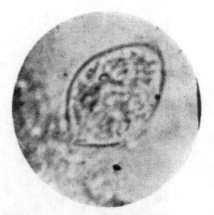

Figure 19. Oval, organized cluster of vesicles, immotile. Film preparation No. 8, magnification 1500×

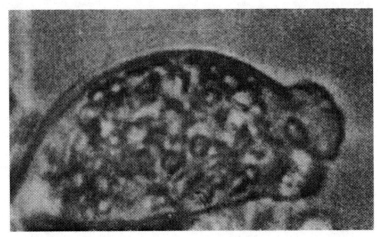

Figure 22. Org-animalcule on edge of blade of grass. Vesicular structure of the protoplasm. 3000×.

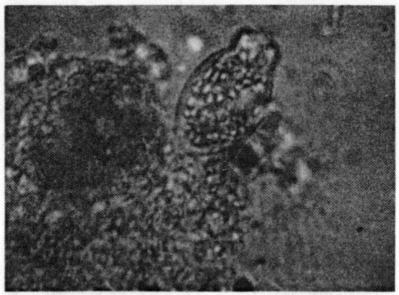

Figure 23. Completely organized vesicular cluster attached to blade of grass. Film preparation No. 8

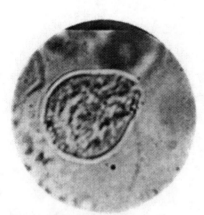

Figure 24. Org-animalcule in ovoid form, 1500×

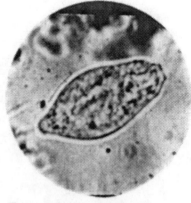

Figure 25. The same organism, stretched

the individual vesicles flowed into its "mouth." This may have been an electrical phenomenon; the animalcules and also the vesicles proved to be negatively charged. The photographs here illustrate various stages of the development of these perhaps identifiable organisms. Spherical or elongated forms can be seen. In addition, we see the same objects hanging by stalks on plant fibers. Some of these plant fibers have clearly broken down into a granular and vesicular structure (Figs. 17–25).

I regard the rhythmic change from spherical form to elongated form as particularly important. Without hesitation one can interpret the stretching as a result of swelling, during which the surface becomes charged. This is the direction "away from the self, toward the world." The intake of fluid and vesicles through the open mouth probably sets in motion a process that leads back to the spherical form through contraction. This contraction can probably be regarded as a discharge process, in the same way that muscles contract when electrically stimulated. If a weak current is passed through the preparation, the contraction takes place prematurely, before the elongation is completed.

Although the detailed relationships are still unclear, this much can be said: In order to be able to proceed further, it was necessary to assume that what I was observing directly was the tension → charge → discharge → relaxation process. Naturally, some crucial questions remained unanswered: for example, what enables this organism to retain its elongated shape and what makes it lose its ability to return to the spherical shape. However, this observation proves that *at least this type of protozoan forms by transformation from plant tissue after passing through a number of developmental phases*. It would not be profitable to attempt to classify these protozoa at this stage. A film was made of the preparation that was studied (preparation 8). The contracting objects were given the name *"Org-Tierchen"* \* (derived from "*org*astic" contraction).

---

\* Org-animalcules. Referred to as org-protozoa in later publications.

## CONFIRMATION OF THE VESICULAR CHARACTER OF THE FLOWING AMOEBAE

The initial assumption that the flowing amoebae are agglomerates of vesicles occurred to me about two years ago. However, the study of earth bions at first led away from the amoeba problem, and this assumption could not be confirmed; no systematic microscopic observations had been carried out. I then pinned my hopes on time-lapse photography to disclose the process on film; however, technical difficulties prevented the problem from being solved for another year. Time-lapse cinematography requires a completely still preparation with no flow whatsoever in the liquid. It was soon found that preparations of grass embedded in paraffin die off because of lack of oxygen. But the filming of vesicles being enclosed within a membrane and the subsequent development of amoebae could be one of the main proofs of my theory. The technical problem was solved in the following way:

Two to three carefully separated pieces of moss or grass were placed on a hanging-drop slide and the liquid for the preparation was added. A cover glass was provided with four small wax feet, one at each corner, and placed over the slide in such a way as to prevent formation of air bubbles. The two long sides of the slide were sealed off with paraffin. It should be noted that about one quarter of the concavity of the slide was not covered over by the cover glass. Water was then deposited along the two unsealed sides to form reservoirs from which the evaporating liquid could be continuously replaced. These two reservoirs of water had to be replenished about once every two hours. Thus, the technical problem was solved and time-lapse filming of the preparation could be carried out over a period of several days. Nevertheless, a certain amount of effort and attention was needed to ensure that the water reservoirs were replenished in time and that the preparation did not dry out. Using these technical methods, it was discovered that, under conditions still not completely understood, dried moss when placed in water generates amoebae in addition to other protozoa. Thin pieces of grass swelled by cooking break down into very fine vesicles. These increase in size and are encircled by a well-defined membrane. Once they have reached a certain size, one or the other portion of the membrane begins to bulge although the whole structure remains motionless. A small circle forms in the center. Then a loose limiting layer forms around the large circumscribed vesicles which have a small, almost dotlike vesicle in the center. This outer margin later becomes the ectoplasm, while the vesicles in the center of the large round saccule become the coarsely vesicular structure of the endoplasm.

After several hours, the object detaches itself from the fiber and crawls away as an amoeba. A flowing motion within the bulging portion of the membrane does not begin until the very last phase of the process. The detachment takes place gradually. Once the formation has freed itself from the plant fragment, one can already see the bulging masses of plasma at various points on the organism. Amoebae are formed continually in this way. Sometimes two or three form simultaneously within a short period of time; sometimes they form one after the other at certain intervals. The amoebae, once detached, undergo division. They can often be observed grouped together in colonies and are difficult to distinguish from the margins of moss fibers that have undergone swelling. Direct observation proved that amoebae form from swelling fibers of moss (see the micrographs from time-lapse film No 8: Figures 26–31).

## MOTILE VESICULAR EARTH CRYSTALS AND EARTH BIONS

My observations and the resulting hypotheses clashed severely with the "germ theory." I deliberately avoided considering the contemporary views on "the origin of life from life germs." Following these initial observations, it was more than ever necessary to approach my work in an unbiased manner. For the sake of comparison, I made several non-sterile water preparations of a tulip leaf, a rose petal, grass, and simple earth. After three days, the tulip leaf and rose petal failed to show any development of protozoa. The grass infusion, on the other hand, was full of motile rods, vesicles, and various protozoa. The earth-water preparation (preparation 1a) contained some major surprises. Microscopic observations of the infusion immediately after it was prepared showed completely motionless, sharply defined, unstructured crystals (Fig. 32). There was also the occasional vesicular formation quivering between the crystals and some elongated, luminescent green, anucleate rods moving slowly. On the third day, the same preparation, which had remained uncovered, looked different. It was full of motile angular structures moving in exactly the same manner as the rods and vesicles. I was struck by the way in which the small vesicular formations were often attached to the surface of rods or larger objects. The formations had a taut, vesiclelike appearance and many of them had modified their structure. *Inside, striations could be found which had broken down here and there into vesicles.* The vesicular inclusions in the structured crystals were indistinguishable from the vesicular formations floating free and tremulous in the liquid.

On the seventh day, both disintegration into vesicles and the structuring had reached a very advanced stage. Even at a magnification of 700× it was possible to see sharply defined protrusions at the boundary of the angular, irregular, brownish-colored lumps of earth. These protrusions looked like *vesicularly structured tubes*, alternately expanding and contracting; bending movements were also observed. At a magnification of 1625×, I saw a brown clump of earth with vesicular protrusions at various points around its edges. It was linked with another clump of earth by a vesicular-striated mass. *At the connection point the clump of earth was bending and stretching.* At first I thought that I was mistaken, but further careful observation left no doubt: *the clump of earth was moving as if it were jointed; it*

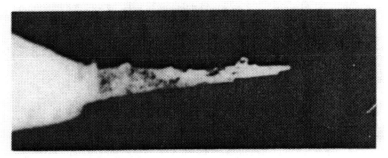

Figure 26. Moss, starting to swell. 300×

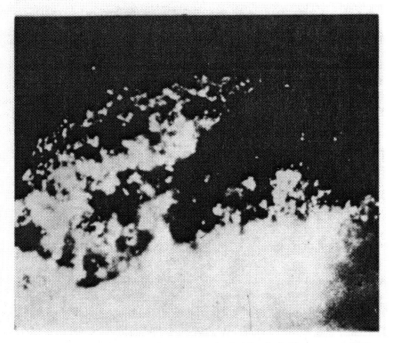

Figure 27. Moss, distintegrating vesicularly. Dark-field, approx. 800×

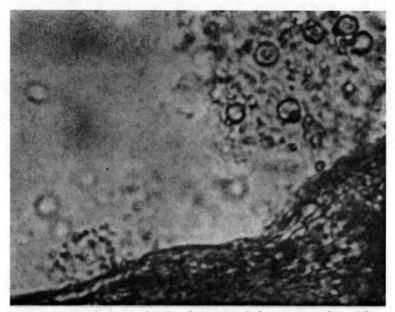

Figure 28. A phase in the development of flowing amoebae. The vesicles at top right are swollen moss and each will develop into an amoeba. Bottom left, another protozoan forming. 1000×

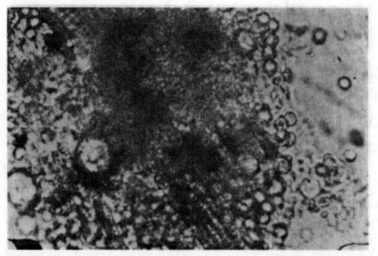

Figure 29. A more mature phase of amoeban development in the same preparation. The vesicles at the left are on the point of flowing. 1000×

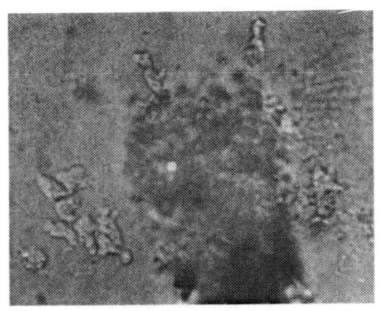

Figure 30. Complete amoebae, moving. Top right, flowing vesicles. 800×

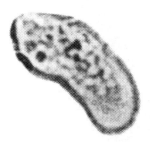

Figure 31. Vesicular structure of an amoeba. 1500×

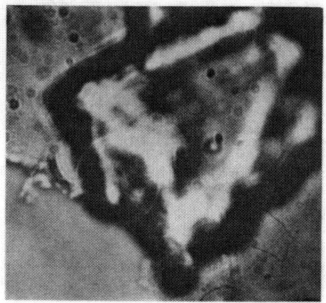

Figure 32. Unstructured crystal of earth. Film preparations 1, 2, and 3. 800×

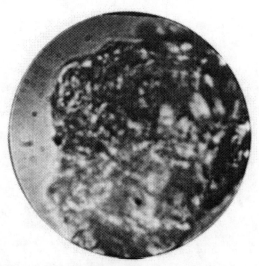

Figure 33. Crystal in early stages of vesicular structuring. Preparation 1a, six weeks old. 1000×

*was stretching and contracting.* After another seven days the process of disintegration into vesicles, the formation of the striated structure, and the breakdown of the edges of the crystals into vesicles had advanced considerably. The protrusions along the edge of the crystal were moving in three different ways: (1) rotating around their axis: (2) stretching and contracting; (3) bending.

I have called these new formations on the crystals "plasmoids" (Figs. 36 and 37).

I passed a current of 1 mA through the preparation, very gradually increasing the strength to 2 mA. The quivering motion of the vesicles accelerated and the stretching and bending became more pronounced. When the current was applied, the vesicles moved toward the *cathode.* They were, therefore, *positively charged particles.* Interrupting the current caused the movement to stop just as promptly as it had started. When the current was reversed, the direction of motion of the individual formations quickly changed. If the current was applied for a long period of time, the swollen tubular structures at the edge of the crystals started to jerk as though they were trying to detach themselves. When the current was interrupted, these detaching motions occasionally continued for a short while. The tubular structures often elongated to twice their original size. Repeated examination of the electrical reaction always yielded the same result: *the passage of current had an as yet undefined influence on the spontaneous motion, particularly on the stretching.*

I shall return to the reactions that I have just described, but now I shall discuss other phenomena that I observed in the further course of my work.

Let us remember that we have so far witnessed two, probably radically different, phenomena:

1. The *vesicular disintegration of swelling plant fibers*; i.e., of *organic tissue.*
2. The *vesicular structuring of crystals of earth*; i.e., of *inorganic*

*matter,* followed by the formation of *motile tubular structures and other moving particles.*

From now on, I repeatedly tried to reproduce the two phenomena experimentally and to randomize the experimental conditions. My primary concern was that my hypothesis was correct; namely, that the vesicles that formed from the swelling substances were in fact identical with the vesicular formations within the amoebae.

I replaced the water in the preparations with 0.1 N potassium chloride in order to test *the swelling effect of the potassium.* It was found that earth treated with potassium chloride swelled more quickly and that the described phenomena were more distinct. The same observations were made on completely dry or semi-dry grass preparations that had been treated with potassium chloride. In the case of the grass preparations, it was possible regularly to produce the immobile clusters of vesicles. More and more frequently, I observed individual vesicles flowing in and out. Similarly, the progressive breakdown into vesicles was reproducible over and over again in the preparations of soil treated with the solution of potassium chloride.

I wanted to make the vesicles, formed during disintegration of the various substances, coalesce artificially by using some suitable agent. I therefore added some very diluted red gelatin to the preparations of grass and soil that were in an advanced stage of decomposition. *Amoeboid structures, which previously had not been present and which had not formed in the preparations without gelatin,* were observed in the various preparations after just a few hours and were completely developed after one to two days. These structures, which I called *"pseudo-amoebae,"* crawled around in *various* directions in the preparations with *jerky* movements. They formed clusters of vesicles, with occasional single rods protruding from their edges like moving spines. The jerky movements were not as organically flowing as the plasma currents in genuine flowing amoebae. At a magnification of 2000×, not only locomotion but also movement *within* the

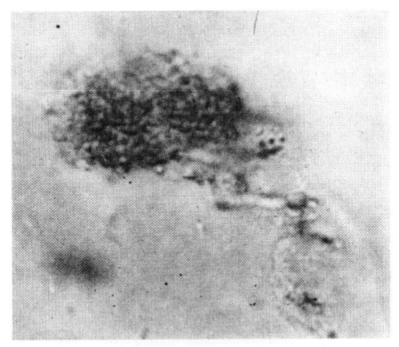

Figure 34. Crawling, vesicularly structured crystal

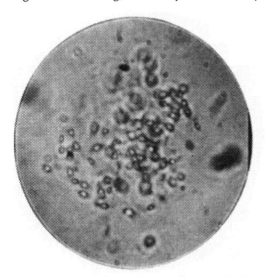

Figure 35. Motile pseudo-amoeba. Sterile earth bions combined together by gelatin

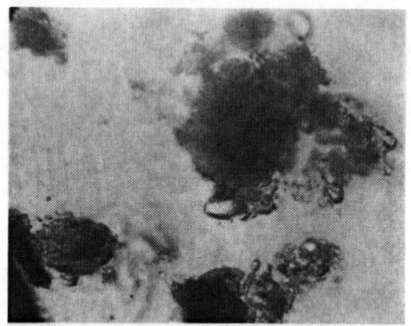

Figure 36. Plasmoidal crystal of earth, motile margins. 1650×

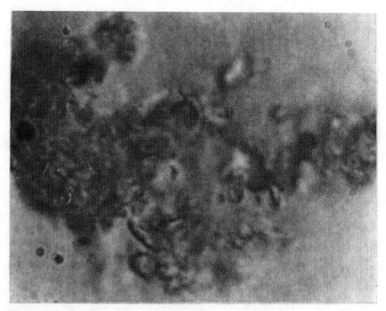

Figure 37. Plasmoidally motile crystal of earth. 2300×

structures and bending and stretching movements were observed. *Completely in keeping with my assumption, the gelatin had thus combined a number of vesicles together into a cluster.* This cluster of vesicles now moved about as a *cohesive whole*. The next problem was to establish the origin of this movement. The plasmatic streaming, seen in many amoebae, was not observed in these preparations (Fig. 35).

Figure 34 shows a completely vesicular-structured crystal connected by a stalk with a second crystal, only the left-hand edge of which was structured. When the micrograph was taken, this crystal was crawling quite slowly in the direction of the arrow and was dragging the incompletely structured crystal behind it.

It seemed logical to object that the pseudo-amoebae described were not at all the result of an artificial coalescence but a result of "germs of organisms" which had infiltrated into the unboiled and uncovered preparation. In order to test the accuracy of this objection, I adopted the practice of boiling the preparations for fifteen to thirty minutes in closed glass containers.

This experiment yielded a completely unexpected result: *the boiled preparations immediately exhibited richer and more active forms of life than did the unboiled preparations after days of swelling.*

In the unboiled preparations, the individual vesicles were scattered and most of the crystals were unstructured. The vesicular structuring of the crystals proceeded *slowly* over a period of several weeks. In contrast, *immediately after boiling,* the preparations exhibited innumerable, constantly moving, round or irregularly shaped vesicles grouped closely together. The liquid of the unboiled preparations usually remained clear and some material settled out. The liquid of the boiled preparations immediately became *colloidally opaque.* Examination of the electrical characteristics of the boiled preparation showed that the vesicles were also *positively charged* particles, which migrated toward the cathode when a current of 0.5 and 1 mA was passed through them. When diluted sterile gelatin was later added, the result obtained was the same as that described earlier for the unboiled soil: *the individual granular vesicles combined to form moving amoeboid structures.* The same result was achieved by boiling a mixture of earth, potassium chloride, and gelatin.

However, the presence of motile forms in a tightly sealed preparation immediately following boiling presented a major problem. The possibility that the boiled preparations could contain much more life than the unboiled, non-sterile, uncovered preparations seemed to go against all the laws of sterilization. Six days after boiling, a sealed preparation showed that, at a magnification of 1500×, the vast majority of the structures exhibited the familiar movements. The preparation also contained some plasmatic projections and formations; i.e., areas which showed up brightly and had three to four thread- or rod-shaped projections.

Most crystals had a vesicular structure throughout. Controls carried out on *fresh, unboiled* earth-water preparations always yielded the same result: the lack of a vesicular structure and fewer anucleate vesicles. In making these observations, I had to learn to distinguish between the spontaneous movement of clumps of soil that had undergone swelling and the passive movements of clumps of soil caused by moving objects colliding with them. There no longer existed any doubt that the boiled preparations contained much more active forms of life. Similarly, there was no doubt that the boiled preparations contained far more abundant forms of contractile tubes and marginal swellings.

When using strong water-immersion lenses at a magnification of 2300 to 3000×, I was able clearly to observe *pulsation* at individual points in the contractile formations. Solid crystalline objects with organized boundaries, seemingly linked to each other by a gelatinous mass, were trying to "free themselves." In some forms, although only at very high magnifications, I was surprised to note that there were green, bright-looking structures moving very vigorously within the gel-like mass. These structures were in all respects identical with the green, bright nucleated rods floating freely around in the liquid. I was forced to assume

## Bions as the Preliminary Stages of Life 51

that the clumps of earth had swelled and that *individual parts had formed into vesicles or rods*. Whether or not these swollen units remain within the crystal or whether they free themselves from it and move around unimpeded in the liquid depends on a number of still unknown factors. When I passed a current of 0.5–1 mA through these preparations, I again observed the formation of vesicular protrusions, intensified quivering of the vesicles, migration toward the cathode, etc.

Thus, it became increasingly clear to me that *the more vesicular the structure of such a crystal, the more motile it is*. In order to continue my work, it was necessary to assume that the vesicular particles were *swollen, electrically highly charged units of matter*, held together by a gel-like substance. These particles move around inside the amoeboid structures just as they would move individually as anucleate vesicles in a free, unrestricted space.

Repeated control experiments with boiled earth always yielded the same phenomena; i.e., motile vesicles, nucleated and anucleate formations, pseudo-amoebae and divisions. In order to ascertain whether these phenomena were not perhaps ascribable to errors in the preparation, I made control preparations of unboiled earth, leaves from trees and tulips, etc., and I left these preparations *unsealed*. Again, *the non-sterile preparations exhibited far fewer vesicles than the boiled preparations and it was much more difficult for motile vesicles to form or for the marginal layers to break down into vesicles than was the case in the boiled preparations.*

The question then became: if these formations were really living matter, did they exhibit any other known properties, such as cell division, in addition to motility, contraction, and expansion? I began to study the earth-potassium chloride-gelatin preparations in detail with this question in mind. After observing one and the same field or one and the same structure under the microscope for several hours, I was soon able to detect *divisions of the pseudo-amoebae*.

As I grew more familiar with the preparations, I became

more and more convinced that these were probably living forms which were, so to speak, incomplete; for example, the movements of the pseudo-amoebae were abrupt, slow, tremulous without any inner flow; i.e., "mechanical." The objects must therefore be in *preliminary stages of life*. For my own convenience, I called them *"bions."* Could complete living organisms develop out of the bions?

I was pleased to note that in addition to division they also exhibited the function of "eating." I watched a *pseudo-amoeboid cluster of vesicles of the type described as it ingested freely swimming individual vesicles*. I had to rid myself of the preconception that is wrongly associated with the word "eating." It is certainly a mistake to use anthropomorphic expressions when describing protozoic formations. When we talk of "eating," we automatically think of a rational living creature "ingesting food from its environment in order to stay alive." I had to rid myself of this false conception before I could include the phenomenon of "eating" along with the other observations in a way that made sense. It seemed irrefutable that *the cells I was observing were composed of vesicles and rods*, in the same way that a body of a multicellular organism is made up of individual cells. This conception clashed with the view that cells are the ultimate units of living matter, for it would mean that not the cells but the vesicles are the biological units and that the cells are already complex structures. As the electrical reaction showed, these vesicles and rods were highly charged, tautly stretched structures. It seemed logical to assume that a cluster of vesicles *"eats" by attracting individual free-swimming and electrically charged vesicles toward itself, adding them to the growing cluster*. This is exactly how contractile protozoa behave.

This was just one assumption that suggested itself. I would have been equally willing to accept any other plausible one. At this stage of the work, it was most important to refine and control the experiments. *I was continually faced with the question of how it was possible for boiled substances to contain more life than unboiled, non-sterile substances.* I originally thought that the rapid quivering motion of the boiled vesicles was a heat phe-

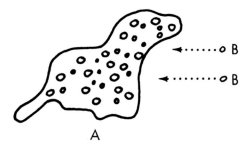

Illustration of the "act of eating"; drawn from life

Amoeba A did not move toward bions B. Instead, the bions, once they had approached to within a certain range, moved quickly toward the amoeba and disappeared inside it (see direction of arrow).

nomenon, but several experiments in which electric current was applied demonstrated that this motion was electrical in character, because the quivering increased when the current was applied. At this stage, I did not want to consider the problem of whether these were true bacteria or cocci or whether other types of structures *had formed in the boiling process.* If it had proved possible to create pseudo-amoebae out of boiled earth, gelatin, and potassium chloride, it should also be possible to carry out a further control experiment with other substances. I felt that there must be substances which would eliminate the "unfinished" character of the formations.

Later I changed over to autoclaving the earth in potassium chloride at 120°C, with a much better result than that achieved with charcoal heated to 180°C. Autoclaving yielded only a few unstructured crystals and almost exclusively motile vesicles and clusters of vesicles. I will come back to this when I describe the culture experiments.

The observations described so far confirmed, to my mind, the biological accuracy of the tension-charge formula. *The vesicular swelling represented the mechanical tension, while the posi-*

tive or negative electrical behavior of the individual vesicles represented the charge. However, large areas of the problem were still obscure.

## EGG-WHITE PREPARATIONS
(Preparation 6)

According to the tension-charge formula, completing the development of function in the structures required certain substances which had the property of swelling. Experiments carried out elsewhere show that lecithin acts like potassium, and cholesterin acts like calcium, the first two having a *swelling* effect and the latter two a *shrinking* effect. Therefore, I adopted the practice of adding lecithin and cholesterin to the soil preparations. In addition to the phenomena already described, others occurred which I had not previously seen. At first I was struck by the fact that the preparation was full of tubelike, regularly formed structures which slowly changed shape (Figs. 38, 39, and 40).

Control experiments, in which each substance was separately given the same treatment, soon showed that these structures were created by the lecithin. They exhibited the following characteristics: the tubelike, often regularly notched, structures became longer and thicker, bent, changed position, and in particular exhibited budding. At certain points on the tube, another tube formed, which in turn branched into another, and so on. If I added KCl to lecithin, I obtained exactly the same formations. *They did not exhibit any organic movement.* I believed that the "growing" and sprouting that I observed could be ascribed to the intake of liquid, as into a sac. This process ceased after about forty-eight hours; however, *there was no doubt that lecithin was capable of creating structures.* The movement connected with this growing and sprouting was fundamentally different from that of the pseudo-amoebae.

*I now added egg white to the earth-lecithin preparation.* The result exceeded all expectations. After a few minutes, round

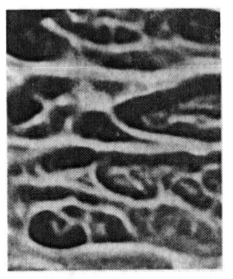

Figure 38. Lecithin, rubbed dry on cover glass. 400×

Figure 39. Cholesterin crystals. 400×

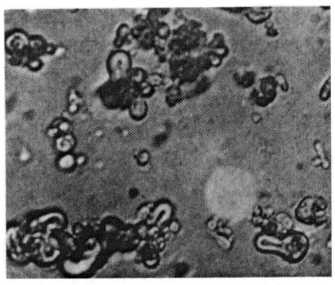

Figure 40. Lecithin tubes, lecithin in 0.1 N potassium chloride. 1200×

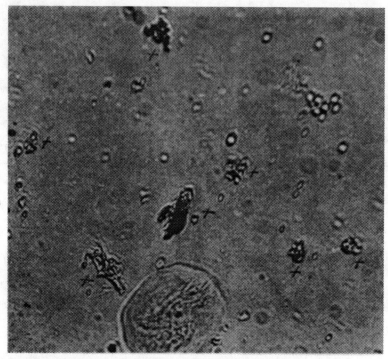

Figure 41. Bions immediately after preparation. + = structures formed through the addition of coal. Bion preparation 6b. 1000×

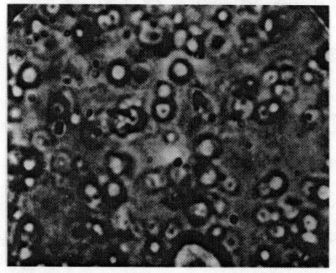

Figure 42. Fresh, sterile bions, bion preparation 6c. Film preparation 6. 2000×

## Bions as the Preliminary Stages of Life 57

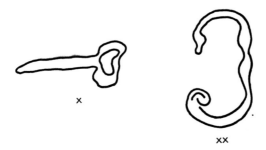

Lecithin structures (drawn)

x  Elongating tube
xx  Greenish shimmering vesicle expanding and contracting. When a current is passed through it, the vesicle stretches toward the anode

cells, which had dark nuclei and underwent frequent division in rapid succession, formed in the unboiled preparations. The cells divided either by constriction in the middle, in the course of which the nucleus also divided, or by the sprouting of a smaller cell, which then separated itself from the larger cell. Detaching motions were clearly observed at certain points. What I am trying to describe succinctly here was actually revealed only after long periods of painstaking observation. To the same extent, as I learned to observe and follow the phenomena, they acquired a lawfulness and I gained confidence in identifying them. When egg white was added, cells formed. *No cells were formed when lecithin and cholesterin were not added. In this case, the preparation contained only agglomerations of flocculate egg white such as can be observed when egg white is placed unboiled under the microscope. Lecithin, with only water added, also failed to produce any motile cells.*

These mixed egg-white preparations also showed that the unboiled preparations formed motile substances more slowly, and were less richly varied than the boiled preparations. The unboiled preparations were much less turbid and contained more sediment. The colloidal particles which formed upon boiling were

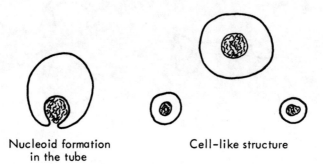

Nucleoid formation
in the tube

Cell-like structure

Cell-like structures formed after the addition of egg white

thus in suspension due, according to all previous knowledge, to their electrical charge. After allowing the preparations to stand for several weeks, I noted an enormous increase in the number of individual formations. After six to eight weeks, the liquid again began to clear and the particles sank to the bottom, forming a thick whitish layer. Microscopic examination revealed that, when the colloid went out of suspension, the motility and the electric

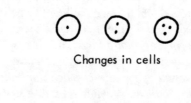

Changes in cells

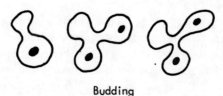

Budding

## Bions as the Preliminary Stages of Life

reaction also ceased; the structures were "dead." Let me briefly recapitulate:

*Lecithin with potassium chloride alone does not produce cells, but only regularly formed tubes of various kinds. There is also no organic movement, but merely a growing and sprouting, apparently caused by the intake of fluid.* Egg white to which only KCl is added does not result in any cell formation. However, *egg white plus lecithin plus KCl plus cholesterin gives rise to cell formation.*

I made the experiment even more complex by adding gelatin. Thus, I had a mixture of egg white, earth, lecithin, cholesterin, gelatin, and potassium chloride. When gelatin was added, vigorously moving plasmoidal vesicular structures were visible immediately after boiling. In contrast to the pseudo-amoebae of the soil preparation described earlier, these objects had a *much finer structure* and displayed much *more rhythmical motion,* no longer jerky and mechanical but smooth and organic (Fig. 41). The amoeboid forms which I obtained in this way grew more complex the longer the preparation was boiled. They were endowed with far more functions than boiled-earth preparations. While the pseudo-amoebae in the soil preparations were capable of movement, the particles within the structures exhibited little if any movement, and only rarely and weakly displayed flowing, bending, and stretching. The amoeboid forms in the mixed egg-white preparations exhibited streaming, internal movement, bending, stretching, and locomotion. The functions of division and "eating" were present in both (Figs. 42 and 43).

The amoeboid forms in the mixed egg-white preparations had a much finer, vesicular structure. In addition, after a few days it was possible to observe formations that possessed homogeneous plasma and a flowing pseudopodal movement. *Contraction, expansion, division, budding, "eating," and locomotion were also visible.* The questions of *metabolism, culturability, stainability, and the bacteriological characteristics* of the structures that had arisen still remained unanswered.

The biological nature, structure, and activity of the motile forms can be altered by using different substances or by adding or omitting one or another substance. This fact must be regarded as important proof of the possibility of life forming out of inanimate matter. The substances do not have to be mixed in exact proportions. It is as if the structures were formed in accordance with some as yet unknown law; as if the relationship of the substances which form living, motile structures was determined by a *self-governing principle*. However, nothing is known about all this.

In my further investigations I added meat broth, some milk, and some egg yolk as nutrient substances to the bion mixture. I also added carbon in various forms: i.e., finely pulverized, dry-sterilized coal; coal dust autoclaved in KCl; coal dust heated to incandescence, dry or in broth; finally, dry-sterilized soot or soot heated to incandescence. The best results were obtained by adding soot: very small but extremely motile, finely structured amoeboid objects formed. Bion mixture 6 is now prepared in the following ways:

6a. Non-sterile mixture.
6ab and 6ac. The non-sterile mixture is boiled (100°C, ½ hour) or autoclaved (120°C, ½–1 hour).
6c. The substances are pre-sterilized.
6cb. The mixture is prepared under sterile conditions and then boiled.
6cc. The sterile mixture is autoclaved once more.
6ccc. The sterile mixture is autoclaved twice.

The 6cc mixture gives the best culture results. The first inoculations are carried out after three to four days.

Is it possible to talk here of the *creation of artificial life?* Out of vanity, I would say yes; by reasoning correctly, I must say no. If, in metaphysical terms, life was a realm completely separate from non-life, I would be entitled to say that I was creating "artificial life." But my experiments showed that there are *developmental stages* in the progression from lifelessness and immobility to life and that, in nature, life is being created out of inorganic matter presumably by the hour and by the minute. In this case, one cannot really talk of artificial life. *I merely succeeded in revealing, experimentally, the developmental process of life.* It was possible that a new form of organism was artificially generated in the process.

On January 8, 1937, I sent Professor du Teil the following report:

Before giving you a detailed report of my experiments I should like first of all to tell you something about a heating experiment based on the tension-charge formula. Here I shall merely mention the experimental procedure and the result. A film is being made at the Institute for Sex-Economic Research which will clearly show what happens. At the same time, samples of the colloidal preparation will be sent off.

In order to lay the groundwork for further experiments, and in the light of the large number of findings already available, it was necessary to regard the formula of *mechanical tension → electrical charge → electrical discharge → mechanical relaxation* [*] *as identical with the formula of vegetative functioning in general.* The following experiment should be carried out to provide proof of the formula:

At this point in my experiments I start by preparing a mixture of 100 cc sterile *Ringer's solution* and 100 cc 0.1N *potassium-chloride solution*. A solution of *red gelatin* is added to this mixture until the latter turns pale pink. A tiny amount of *coal dust* picked up with a tweezer and an equal quantity of cholesterin crystals are added. The entire experiment is based on the principle of bringing together in a certain sequence those substances that are needed for cell synthesis.

We now dissolve a teaspoonful of fresh, clear *egg white* in about 50 cc of sterile potassium-chloride solution and add both to the previous mixture. The egg white dissolves after being

[*] See "Der Urgegensatz des vegetativen Lebens" (1934) and "Experimentelle Ergebnisse über die elektrische Funktion von Sexualität und Angst" (1937), both published by the Sexpol-Verlag, Oslo.

stirred for a short while. Under the microscope it is not possible to detect any formation of shapes, nor is there any sign of movement or of plasmatic structures. Even at high magnification (1000×) it is only possible to observe the typical immobile crystals of coal and cholesterin.

We now add 1–2 cc of milk and some egg yolk. The latter makes the previously clear mixture opaque.

Now a *second solution* is made: *lecithin in salve form is triturated in KCl solution.* At a magnification of 500–900×, we see strange objects developing and growing; namely, *tubes* that swell, bud, and bend. There is no discernible structure inside the tubes, although occasionally there are clusters of vesicles of definite shape. No organic movement occurs. These are purely physical swelling phenomena based on changes in the ratio of internal pressure and surface tension.

If we now add the lecithin solution to the first mixture, the latter *immediately and progressively* turns yellowishly-gray opaque. Under the microscope, one is surprised to observe motile life forms: *quivering movement from place to place, buddings, divisions of cell-like formations, jerky to flowing crawling motion.* The organic character of the movement can only be observed at magnifications in excess of 1500×, *but it is most clearly visible at 2300–3000× using a binocular microscope.* If a formation is continuously observed at about 3000×, one notes brightly glimmering, vesicular, nucleated objects which change their shape. It is possible to distinguish four groups of vigorously moving structures: round, nucleus-like *vesicles; rods;* round, *cell-like nucleated forms* that move about but do not display any inner motion; and finally, *amoeboid structures.* These last are particularly interesting because at magnifications of 3000–3500× they reveal movements of contraction and expansion. At only 2000×, one can already see them crawling.

The structures formed in this way are, at least according to the experiments conducted so far, *negatively charged bodies;* they migrate to the anode. If the colloidal mixture is allowed to stand six to eight weeks, a thick sediment forms, and the liquid becomes increasingly clear. When the sediment is examined at the magnifications mentioned above, it is found that the motility that was described has been lost and the objects are now "dead." A report on the control experiments that were carried out will be given in the detailed description. At present, the metabolism and the stainability are being studied. It has been ascertained that *the structures are culturable.* If, in fact, the mixture is boiled down until it no longer contains any liquid and the residue is inoculated with sterile colloidal mixture, growths are produced which so far have been cultured to the fifth generation.

The structures, provided they are kept sterile, have proved to be innocuous in animal experiments when injected subcutaneously in mice and guinea pigs.

The question whether these are complete living forms can only be definitively answered in conjunction with other experiments, which will be reported on later, and only after all experiments and controls have been carried out. The described experiment was captured on film.

The artificial, lifelike structures were given the name *"bions."*

# 3

# The Culturability of the Bions (Preparation 6)

*Bions are preliminary stages of life; they are transitional forms from the inorganic and non-motile to the organic, motile, and culturable state.* In order to propagate bions, whether from earth, coal, animal tissue, moss, or mixtures of these substances, very careful observations are needed to identify the conditions under which bions become culturable. The fact that bions are preliminary stages of life and not "completed" forms of life was further evidenced by the problems encountered when attempts were made to culture them. The decisive factors determining culturability are the material composition of bions and the nutrients contained in the culture media.\*

---

\* *Nutrient medium c:* Ringer's solution and KCl were mixed; then some red gelatin was added until the liquid turned slightly pink. Next, coal, cholesterin, a quarter of the white of an egg, several drops of milk, and a very small amount of egg yolk were added. Finally, some lecithin and a whole egg were added to the complete mixture. This mixture was then divided among test tubes and dried out at 80–100°C for one hour in a horizontal position so that the nutrient medium formed as a flat deposit. A small droplet of condensation remained behind. The nutrient substrate was yellow.

*Nutrient medium d:* Composition and treatment the same as for nutrient medium c. Growth occurred in some test tubes, although no inoculation had been carried out ("auto-inoculation").

*Nutrient medium e:* A very small amount of Ringer's solution and KCl were mixed. Also, a few drops of red gelatin dissolved in KCl were added; coal was triturated and heated in KCl. Also, a tablespoonful of milk and a whole egg plus lecithin were triturated in KCl and added to the entire mixture. This mixture is then divided among the test tubes and evaporated for two hours at 80–100°C in a horizontal position. *This was done for two*

## The Culturability of the Bions (Preparation 6)  65

Since autumn of 1936, fresh bion preparations (6) have been made on an average of two to three times a week in my laboratory. On January 1, 1937, a fresh bion mixture that had been boiled for an hour was inoculated into a nutrient medium. This nutrient medium (a) consisted of the dry residue that was left after the bion substances had been completely boiled down. On January 3, forty-eight hours later, a grayish-white moisture, which continued to build up, formed on the nutrient medium. *The medium was disintegrating.* Both the nutrient medium and the bions had been rendered absolutely sterile by hours of boiling. The nutrient medium in question (a) had stood for weeks in a constant-temperature incubator without showing any signs of decomposition. The first-generation culture seemed to have succeeded. Under the microscope it was possible to make out *all four types* that were seen in the bion mixture: round *vesicular cocci*, short and long *rods*, round nucleated cells, and finally, contractile and expansile, *crawling amoeboid forms.* The *cultivated formations* appeared more motile and vigorous than the originally boiled bion mixture. The records of these tests show that this strain was successfully propagated while varying the preparation of the sterile culture medium, until February 9; that is, to the ninth generation. Propagation was interrupted only once, on January 9, 1937, because the culture medium in question was too dry. The third generation which developed on January 9, 1937, was inoculated onto culture medium (b) on January 14. I obtained culture medium (b) by not quite completely boiling the remaining substance dry. Apart from this one interruption, the propagation experiments ran completely undisturbed.

Within twenty-four or forty-eight hours, the previously dry nutrient medium regularly decomposed into a grayish-white liquid. The liquid was swarming with motile bionlike structures.

---

*days in succession.* No autolysis took place in the test tubes. This was checked by allowing the test tubes to stand for five days at 37°C without any detectable growth.

*Nutrient medium f:* Usual broth and agar nutrient medium.

With the exception of nutrient medium (d), the non-inoculated control media *showed absolutely no signs of decomposition.* If the nutrient medium was not boiled down sufficiently, so that some liquid remained, *"auto-inoculation,"* as I called it, sometimes took place. Despite the long period of boiling, some bions had obviously remained and caused the nutrient medium to decompose. This *auto*-inoculation, or *auto*lysis of the nutrient medium without inoculation, was an undesirable disruption that confused the clear outcome of the culture tests. Nevertheless, the auto-inoculation could not be regarded as a negative sign; if the bions could not be destroyed by cooking and if the slightest traces of moisture in the mixture could produce motile forms of life, then it was conceivable that "the nutrient medium itself was alive." It became essential to not only eliminate auto-inoculation but to find a nutrient medium that did not possess the property of auto-inoculation.

Before I hit upon the idea of boiling the culture medium mixture to dryness and using the residue as a nutrient medium, many weeks had been spent painstakingly attempting to propagate the bions in Ringer's solution, without achieving any clear result. In contrast, the following test was a complete surprise: A culture prepared from the seventh generation of the first strain of the sterile 6b preparation was heated for fifteen minutes in the sterilizer. The heated culture was inoculated on February 8 onto a new variation (e) of the old bion nutrient medium. At the same time, the seventh culture generation was inoculated, one sample heated and one unheated, onto broth plus agar-agar. Only twenty-four hours later, to my extreme surprise, I noted that the heated as well as the unheated cultures of the seventh generation had formed dense clusters on the agar-agar and egg-white media. *The agar medium was not broken up.* The growth of the heated cultures was macroscopically identical with that of the unheated cultures. Although this result might at first appear surprising, it is logical, for if it should ultimately turn out that bions are living formations that cannot be destroyed by heating then it would be only natural for the cultures to produce further

## The Culturability of the Bions (Preparation 6)   67

cultures after heating. The results of this experiment have already been confirmed.

On January 9, 1937, a *second,* fresh, two-day-old sterile strain of bions was inoculated onto nutrient medium (e). On January 11, the nutrient medium in *one* test tube had decomposed as in the case of strain I. The control medium had not disintegrated, but a second inoculated medium had not disintegrated either.

At that stage of my investigations, a first generation of a bion strain often did not grow on the nutrient media. On January 22, after unsuccessful culture tests on nutrient media (a) and (b), *the strain that did "take" was inoculated successfully onto nutrient media (e).* There was also clear evidence of disintegration in two test tubes. Some of this second culture generation of strain II was inoculated on January 22 onto nutrient medium (e) and this resulted in new growth the next day. On January 27, *a heated culture generation* of strain II was inoculated onto the same nutrient medium (e) and this resulted in the *same type* of growth on February 4. *The heated culture test was thus successful in the case of strain II, just as it was in the case of strain I.* We continued growing the unheated culture until February 9, always with positive results, as indicated in the table. Meanwhile, the heated culture generation II b/2 had been reinoculated onto culture medium (e) on February 8. Within twenty-four hours, growth in the form of grayish-white, turbid, aqueous disintegration of the nutrient medium was detected (II b/3).

Similar success was achieved with *a third strain (III) of bions* inoculated onto culture medium (d) on January 21, 1937.

I would now like to describe some of the basic phenomena that must be watched for when assessing successful and unsuccessful cultures:

1. After forty-eight hours, but sometimes after much longer, a moist, glistening, uniform coating forms on the nutrient medium in the case of a successful culture. When the control nutrient medium is examined carefully with the aid of a

magnifying glass, the hummocks and the very shiny appearance of the coating are missing.
2. When no disintegration takes place, the surface of the control medium remains smooth. In contrast, usually rounded, but occasionally elongated, grayish-white bumps form on the inoculated medium.
3. Culture medium (d) clearly gives off condensation. The condensation of the control nutrient medium usually dries out in the incubator, while in the case of the inoculated medium the liquid that is given off renders the medium turbid and causes deep cracks to form.
4. Occasionally, reddish growth, which has not yet been explained, was observed.

The fact that the later generations yield new growth much more easily than entirely fresh mixtures is very important. The later generations also display more vigorous movement and have a more complicated structure when observed at 3700× under a binocular microscope; one might say that they appear more "finished." The four basic forms (round vesicles, long motile rods, nucleated cells, and amoeboid crawling forms) appear everywhere. But when they are observed very closely for a long time at 3000–4000×, even the types themselves are not uniform but appear to be subdivided again into subspecies. I will report more fully on this on another occasion.

After many successful, but not entirely unambiguous, culture tests on egg white–yolk nutrient media, I was able to culture bions on broth and agar in the following manner. Bion mixture 6 was heated for half an hour in test tubes in the dry sterilizer set at 160°C. The heated mixture was stored under sterile conditions for two to four days and then inoculated onto broth and agar nutrient media as follows: after the pipette as well as the mouths of the test tubes were flamed, a drop of the mixture was transferred to the latter. The first broth inoculations from three different bion mixtures invariably yielded a heavy turbidity after twenty-four hours.

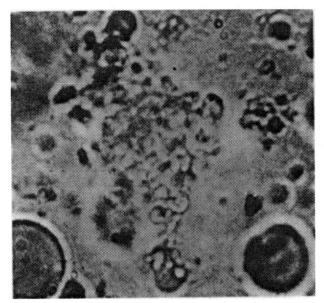

Figure 43. Vibrating "amoeba" moving from place to place. Bion preparation 6. 2300×

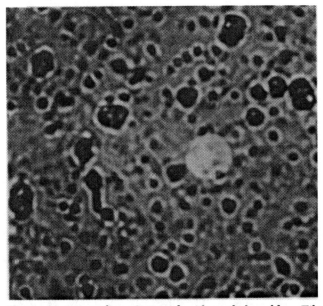

Figure 44. Bion culture 6 stained with methylene blue. Film preparation 6. 1650×

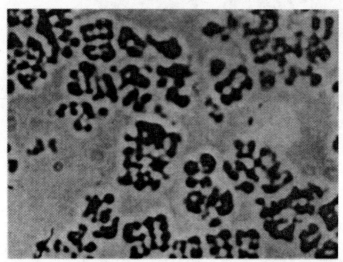

Figure 45. "Packet amoebae." Culture 6cI of the first autoclaved bions 6. Film preparation 6. 2000×

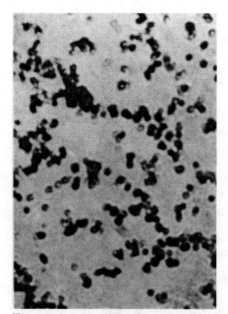

Figure 46. Packet amoeba culture 6cI, gram-stained. 1000×

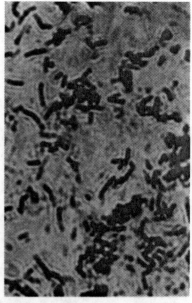

Figure 47. Bion culture 6cbIII, gram-stained. 1000×

## The Culturability of the Bions (Preparation 6)  71

The following features were detectable macroscopically in the broth cultures: cloudy turbidity, flakes, and a fine, thread-like rising cloud of particles, which increased the turbidity when the test tubes were shaken.

Growths on the agar media were macroscopically less clear. Sometimes a very dense yellowish-white growth formed after only twenty-four hours, but in most cases merely a very thin bumpy growth of the same color as its surroundings resulted, becoming somewhat denser after a few days.

Under the microscope, all the broth cultures exhibited rapidly moving rods of various types, also cocci, but very few amoeboid forms. The dense agar growth consisted largely of cocci moving very vigorously from place to place. In contrast, the weak agar growth exhibited much less movement when examined under the microscope. If, however, the weak second-generation agar growth was propagated in broth, the same formations were obtained and the broth culture took on the same macroscopic appearance as in the case of direct inoculation into broth. The nutrient medium thus has an effect on the type of culture. If inoculation was carried out from broth culture to agar, a dense growth occurred regularly within twenty-four hours. Strangely, when the inoculation was carried out from the broth to agar, the cocci predominated in the agar growth, whereas when inoculation was carried out back to the broth suddenly the rods and amoeboid forms again predominated.

In order to make the culture experiments completely unambiguous, I proceeded to cultivate *autoclaved* bion mixtures (6ac). The following experiment was performed: The bion mixture was prepared as described in the first report, the sole difference being that the coal was first finely pulverized, then cooked in KCl, and then added to the first mixture. Two test tubes were filled with the solution and put in the autoclave. The autoclave was set at 120°C and the test tubes remained in it for a half hour. After twenty-four hours, inoculations were carried out from one of the test tubes onto a broth medium and an agar medium. Both resulted in growth which again was much stronger in the broth than on the agar. The first broth culture generation of the autoclaved bion mixture was continued in broth and agar, also egg medium and blood agar. Each case resulted in strong growth that contained moving rods, cocci, and amoeboid forms. At the same time, material from the first broth generation was inoculated onto an agar nutrient medium. After only twenty-four hours, a thick yellow growth appeared on the agar medium, which, to my surprise, when viewed under the microscope, turned out to be a pure culture of amoeboid forms. At 3700×, the particles inside the amoebae vibrated vigorously and the amoebae moved along slowly while simultaneously expanding and contracting. The second generation obtained in this way was further inoculated onto agar and this resulted in a dense growth of similar micro- and macroscopic appearance. The pronounced structuring of the amoeboid forms was very striking and, from my point of view, a positive feature. Over a long period of observation, these objects were seen to undergo division repeatedly. This was seen most clearly at about 3000×.

Once the culturability of the bion mixture that had been heated for a half hour had been established and once the first culture of the first autoclaved bion mixture 6cI had succeeded, my assistant took over the task of preparing the bion mixtures and also of carrying out culture experiments. In the laboratory, she was able to obtain cultures of similar type from eight 6cb bion mixtures. At the same time, the 6cb IX, 6cb X, 6cb XI bion mixtures were sent to Professor du Teil in Nice for control inoculations. We were *unable* to cultivate preparation 6cb IX, but preparations 6cb X and 6cb XI produced growth just as promptly as the previous six preparations. When the preparations were sent to Professor du Teil, I wrote that IX would probably not produce growth because it had a poor macroscopic appearance. This suspicion was confirmed by the fact that we were unable to cultivate this particular preparation. Professor du Teil informed me that, whereas preparations 6cb X and 6cb XI had promptly formed growth, preparation IX had not. The following mixtures, from 6cb XII to 6cb XX, did not produce a single culture. We were unable to explain the reason for this.

## The Culturability of the Bions (Preparation 6)

The mixtures had been prepared in the same manner as before, yet now there were no cultures, in contrast to the previous series of cultures. After much thought, we decided to check the electrical properties of the preparations, which was something that had not been done for a long time. It turned out that all the bion mixtures that had yielded cultures were *negatively charged*, whereas *the mixtures that had yielded no cultures were electrically neutral*. Preparation 6cb IX was one of the series of bion mixtures which had been expected to yield cultures but had not done so; in contrast to 6cb XI and 6cb X, it had shown no electrical charge. Thereafter, bion mixtures that did not exhibit any electrical charge were no longer inoculated, so as not to impair the statistics unnecessarily. However, of the series of bion mixtures that had failed, one after the other, to produce any bion cultures, there was one that did produce a culture. In contrast to the others which had been electrically neutral, this one had a negative electrical charge. This seemed to me to be the final proof that *the electrical charge of the mixture is an essential precondition for culturability*.

While the fresh bion mixtures migrated toward the anode, or were neutral, electrical examination of the second generation on agar (yellow "packet amoebae") revealed a very clear tendency of the objects to migrate toward the cathode. The nature and origin of these forms has so far not been clearly determined.

Up to the present date, fifty-three generations of the yellow amoebae [*] (strain 6ac I) have been propagated on agar. When they began to lose their shape and their yellow color, they were inoculated onto blood agar and egg nutrient medium. This restored their original structure and color. We have not yet seen this form appear again in any type-6 mixture experiments (Fig. 45). (The written reports to Professor du Teil provide some information on particular cultivation problems.)

As time went by, I grew more experienced in culturing the

---

[*] According to the views held until now, the expression "amoebae" is not quite correct. It is justified here since it is assumed that true amoebae (limax, etc.) are also nothing more than agglomerations of vesicles.

bions. Many safety measures were taken to ensure that the culture *succeeded* under conditions of complete sterility:

1. The substances are individually autoclaved or dry-sterilized, and inoculated separately into broth. The individual inoculations should not show any growth. This is the sterility control.
2. The substances are mixed as stated. After each addition, a broth is inoculated. This control should not exhibit any growth.
3. The finished mixture is electrically examined. It should be *very highly* charged.
4. The mixture is allowed to stand three to four days in the incubator. It must be uniformly colloidal and exhibit *vigorous motion* under the microscope.
5. Inoculations are carried out from one test tube onto:
    2 broths
    2 egg media (e)
    2 agars and blood agars
6. Inoculations are made from successful broth cultures to agar and egg medium (e).

In this way, many conditions for successful cultivation are met.

## ELECTRICAL EXPERIMENTS

The following detailed studies were carried out by an assistant at the Institute. The preparations in question were produced in the laboratory by the assistants themselves. The report covers only nineteen No. 6-type bion preparations.

So far, the experiments have not yielded any answer as to why the mixture does not have an electrical charge. The only certainty is that the bions must be electrically charged in order to be culturable.

### Experimental Apparatus

The current is supplied via a rectifier with a built-in transformer, a grid, and a resistor. This makes it possible to use any

## The Culturability of the Bions (Preparation 6)

desired current strength between 0–5 mA and 0–50 mA. A current reverser permits the poles to be reversed in order to check the direction of migration more accurately.

A short glass cylinder of 1.5 cm inner diameter and open at both ends was bonded with Canada balsam to the usual slides used in these tests. The balsam is applied hot in a thinly viscous state. After drying, the excess is removed by xylol.

Two bare platinum electrodes pass through the Canada balsam and each projects 0.2 mm into the inside of the vessel thus created. The current is supplied from the rectifier via the platinum wires of an electrode stand; the wires run inside glass tubes.

Dark field microscopic examination is carried out at 400×. A binocular microscope (with inclined tube) manufactured by Reichert ("Z"-Mikroskop) was used.

### General Experimental Arrangement

Unless otherwise stated, the amperage used was 2 mA in all cases. The object to be studied is suspended in physiological salt solution. Some drops of this suspension are introduced into the special slide described, so that the bottom is evenly covered.

Each time, before the current is switched on, it is necessary to check carefully whether there are any flowing movements in the preparation that might be caused by thermal or mechanical events. One should avoid putting too much liquid into the container, because this can also cause very disruptive movement in the fluid.

If there is no sign of directional motion (streamings) under the microscope when the current is switched off, the current is then switched on and it is noted if and to which pole the particles migrate.

When the current is interrupted, a further check is carried out to see whether directional motion can be observed.

Finally, the current is passed through the preparation in the opposite direction and it is noted whether the particles now reverse their direction of migration; i.e., whether they move back toward the same pole as at first, indicating that migration is caused by the current. If frequent repetition of this experiment yields constant results, then the cataphoretic character of the migration is considered confirmed.

It must be stressed that it is not the aim of these studies to make quantitative determinations but merely to determine the direction of migration of the structures examined.

*I. Electrical experiments on individual substances*

1. A small drop of *egg white* in approximately 10 cc of a mixture of equal parts of N/10 KCl and Ringer's solution was autoclaved for a half hour at 120°C (at 2½ atmospheres pressure), then kept in the incubator for forty-eight hours at 37°C.

    The electrical analysis revealed no cataphoretic migration. Only quite weak contractile movements were seen in this preparation. Egg white treated in this way thus exhibits no cataphoresis. Inoculation tests were negative; there were no cultures.

2. On the other hand, *lecithin* treated and studied as described under 1 reacted, even if only weakly, to the passage of current. The typical lecithin shapes and the vesicles formed during autoclaving migrated at low speed *toward the anode*.

    Inoculations from this lecithin preparation into broth and on agar yielded a culture with the same electrical reaction: migration toward the anode.

3. *Cholesterin* was triturated in a porcelain mortar together with KCl and Ringer's solution and then treated as under 1 and 2. No change in structure was observed after autoclavation: the particles did not move in a potential gradient; culture tests were negative.

4. & 5. In undiluted *cow's milk*, which was autoclaved under the same conditions and kept for twenty-four or forty-eight hours in the incubator, and in cow's milk diluted with Ringer's solution and KCl solution, the large number of fat globules moved fairly vigorously *toward the anode*.

## The Culturability of the Bions (Preparation 6)   77

Some culture experiments were positive. The formations in the cultures also migrated *toward the anode*.

6. *Red gelatin*, dissolved in KCl and Ringer's solution, autoclaved, and kept for forty-eight hours in the incubator, behaved as in 1 (egg white): contractile motions in the preparation, no cataphoresis, no culture.

*Remarks on Experiments 1–6.* Only the substances which exhibited cataphoresis after the described treatment occasionally resulted in a culture when inoculated.

The substances with the strongest charge (the highest speed of migration) produced the fastest, strongest, and also most macroscopically clearly detectable growth. The cultures from the preparations with the weaker electrical reaction produced slower and weaker growth.

### II. Examinations of composite substances

7. *Garden earth.* After being autoclaved, most of the earth crystals lost their typical structure and disintegrated into very small vesicles.

    When the current was off, some of these vesicles moved at a fairly high, but erratic, speed across the field of vision.

    Once the current was switched on, they moved almost in single file at low speed *toward the cathode*.

    When the current was reversed, their direction of migration also changed.

    Non-autoclaved particles of earth did not migrate at that stage of disintegration.

    The current was applied continuously for an hour, at a strength of 5 mA. After this galvanization had been carried out, it was found that the particles now moved in a different direction, mainly *toward the anode*.

    Uninterrupted observation for half an hour with the same current strength and direction showed that the particles continued to migrate toward the anode.

    All repetitions of this experiment yielded the same results. The vesicles that form from autoclaved garden earth change their direction of motion after long exposure to a galvanic current of 5 mA: *they first move toward the cathode, later they move toward the anode.* Garden earth from Denmark behaved in the same way in similar experiments.

Cultures were obtained from both types of earth after this treatment.

8. *Garden earth,* dry-sterilized for two and a half hours at 180°C, then heated in KCl, migrated toward the cathode. The particles moved at different speeds but, on the whole, uniformly and consistently.

   This preparation was also galvanized for one hour, but at 10 mA; migration toward the anode was obtained again.

   This experiment was repeated many times and *only once* was a deviating result recorded: the particles moved immediately toward the anode without being galvanized. Since this was an isolated case, it was not possible to say what caused this deviation.

9. Ordinary gas *coke,* even cinders, was pulverized, mixed with KCl and Ringer's solution, autoclaved and held for forty-eight hours at 37°C in the incubator. As a result of this treatment, the coke also broke down, although to a lesser extent, into small vesicles that had the same properties as those developed from earth.

10. "Incandescent coal": pulverized coke was heated on a spatula in a flame to incandescence and then immediately placed in sterile KCl.

    The same independent and cataphoretic movement phenomena as observed in the autoclaved mixture of coal, KCl, and Ringer's solution were also seen in sterile-sampled drops of this mixture.

    When a current was passed through both coal preparations, there was a changeover from cathodic to anodic migration, as had happened in the soil preparation.

All these preparations (8, 9, 10) produced cultures in broth and on agar.

*Remarks on experiments 7–10.* Microscopic observation during

## The Culturability of the Bions (Preparation 6)

galvanization was abandoned because of the formation of gas bubbles at the amperage used (5 and 10 mA). The observation reported in 7 was made on samples which were continuously taken from the galvanized preparation and through which a current of 2 mA was passed while they were under the microscope. The non-homogeneous substances studied here again showed the same relationships between electrical charge, cataphoresis, and culturability as in 1–6. On those occasions when no migration occurred—this happened somewhat more frequently in the case of dust particles, which are particularly non-homogeneous—the culture tests were also negative.

### III. Studies on preparation 6

The following review of the results shows again the correlation between the electrical properties and culturability:

11. 6c I. This (autoclaved) preparation exhibited anodic migration. It was inoculated onto a nutrient medium consisting of substances contained in the preparation that had been stirred into a firm mass after adding egg white. This nutrient medium was first sterilized and then kept in the incubator as a control for forty-eight hours.

    The cultivation test was successful: *two types of growth* formed, one soft and slimy, the other hard. Each culture type was then separately and successfully transferred to agar. Microscopic observation of the soft, slimy culture revealed typical "packet amoebae." The hard culture consisted exclusively of vesicles.

    When an electric current was passed through them, the packet amoebae migrated toward the cathode; i.e., they had a positive charge. By galvanizing them, it was possible to reverse their direction of migration.

    On the other hand, the vesicles, like the small bodies in the preparation, migrated at a uniform speed toward the anode.

    *Note:* The uniform cataphoretic migration is more clearly observable in the cultures than in the original preparation. Since only cataphoretically reacting preparations yielded cultures, it is logical to conclude that the formations in the cultures are derived from those of the preparation which migrate during an applied electrical gradient. The cataphoresis must then be more extensive and consequently must stand out more clearly in the cultures.

12. *6c II.* The preparation also contained the bion formations (vesicles). They moved toward the anode. The culture was fairly hard, with no soft growth, and behaved in a similar cataphoretic manner.
13. & 14. *6c III, 6c IV* exhibited the same characteristics.
15. *Note:* From now on, the following procedure was used. Finely pulverized coal was dry-sterilized at 180°C; lecithin, cholesterin, gelatin, egg white, and egg yolk, as well as KCl and Ringer's solution, were autoclaved. The sterile-prepared mixture of these substances was then heated in the sterilizer at 180°C until it became very turbid.

*6cb V.* A sample taken from the distinctly turbid preparation right after heating immediately exhibited the bion pattern under the microscope.

When the inoculations were taken from this preparation, the storage vessel was opened a few times. When the electrical examination was later made, the preparation proved to be very impure. Apart from the vesicles, there were clusters of vesicles which were strongly reminiscent of fermentation cultures, although it was not possible to say that they occurred in conjunction with fermentation phenomena. Furthermore, the preparation contained a quantity of filamentous bacteria which appeared to be completely lifeless. Apart from the vesicles and clusters of vesicles, slowly and sluggishly moving rods of various sizes were a main component of the preparation.

Most of the structures in this preparation migrated to the anode. However, some rods were clearly seen to move toward the cathode. After a short while, these rods seemed to reverse their charge: they quivered strongly for a moment and moved toward the anode; then they soon became posi-

## The Culturability of the Bions (Preparation 6)    81

tively charged again and moved once more toward the cathode. The long, obviously lifeless filamentous bacteria did not move at all. The culture prepared from the still uncontaminated preparation immediately after it was removed from the incubator consisted of the same microorganisms as the earlier cultures that were prepared from mixtures of the same substances; they exhibited the same cataphoretic phenomena.

16. *6cb VI* presented a very beautiful pattern of vesicles and vesicle clusters. All the vesicles migrated at uniform speed toward the anode, as did the small clusters of vesicles, although more slowly. At this amperage, the large clusters of vesicles did not move at all.

    The inoculation yielded a pure culture consisting of vesicles and clusters of vesicles which moved strongly and clearly toward the anode.

17. & 18. *6cb VII, VIII* presented the same pattern both in the preparation and in the culture.

19. *6cb IX* exhibited the same formations as the preceding preparations but made a much less lively impression and there were no signs of migration in any of the numerous electrical experiments that were conducted. When the current was applied, fairly strong contraction movements sometimes occurred, but there was no cataphoresis, as already stated.

    Cultivation tests were negative: there was no growth even after six weeks.

20. *6cb X.* The preparation was identical in appearance to the preceding ones. The migratory movements were less clear than in 17–19. While the general direction of motion was toward the anode, some individual particles moved perpendicular to the direction of current.

    Cultivation was successful and the cultures had the same properties as those achieved in 17–19.

21. *6cb XI.* Migration was more pronounced. The culture behaved in the same manner as the previous one.

22. & 23. *6cb XII, XIII.* In this case, the egg white was replaced

by autoclaved meat. Large quantities of destroyed muscle fibers could be seen under the microscope.

No cataphoresis occurred. No cultures could be grown.

24.–27. *6cb XIV, XV, XVI, XVII.* In this case, egg white was used once more, and the meat was omitted.

There was no migration and the culture tests were negative.

28. & 29. *6cb XVIII, XIX.* Weak signs of electrical charge—faint cataphoretic phenomena—indicated a more amphoteric character.

No cultures could be grown.

*Remarks regarding experiments 12–29.* When the preparation exhibited no electrical reaction, no culture resulted. If cataphoresis was strongly pronounced, growth usually occurred after twenty-four to forty-eight hours; if the migratory phenomena were weak, then the growth usually did not develop until after eight to ten days. The migratory phenomena of the cultures corresponded to those that predominated in the original preparations.

The results of the still incomplete differentiation tests (biological stain reactions of various types, comparison with various known species of microorganism, interaction effects, cultures of galvanized bions, application of known bacteriological methods to the bion cultures, etc.) will be made known in one of the next reports.

*Postscript regarding the production of bion preparation 6*

The number of successful bion cultures produced from mixture 6 increases considerably if the procedure described in the first report of December 1936 is modified as follows. (This modification was suggested partly by the control tests carried out by Professor Roger du Teil in Nice and partly by further observations made when preparing the blood-charcoal bions.)

1. *Milk* is *no longer* added (*du Teil's findings*).
2. Ringer's solution is replaced by a few cc of *beef broth*.
3. Instead of coal dust or soot, *blood charcoal* heated to incandescence is used.

4. The *first* inoculation is made into *broth* + *KCl* and broth alone.
5. The *second* inoculation is made from the turbid broth onto *fresh* egg nutrient medium. The hummocks which regularly occur are spread out.
6. Growth from the egg medium is inoculated onto blood agar. The growth is *light blue-gray* in color.

# 4

## The Beginning of Control Experiments by Professor Roger du Teil at Centre Universitaire Méditérranéen de Nice

I will now interrupt my descriptions to include my communications to Professor Roger du Teil and also to quote his reports and parts of our correspondence.* This material provides information on important details and on some initial uncertainties that cropped up in January–April 1937. It also helps me to keep my presentation as simple and as tightly knit as possible.

Report given on March 7, 1937
by Prof. Roger du Teil
to the Natural Philosophical Society in Nice
on the work performed by Dr. Reich (Oslo)

Ever since he settled in Oslo, Norway, Dr. Reich, who as a pupil of Freud first specialized in psychoanalysis, has devoted himself to laboratory studies in the field connected with his special subject. Following his discovery of the consistent electrical charge at the surface of the erogenous zones of the human body, he observed the changes in potential that occur in these zones during certain sensations and feelings, in particular during joy, sadness, and fear. For this purpose, he constructed a voltage-measuring device consisting essentially of an electron tube in a circuit which was connected to an oscillograph. The traces generated by the light beam of the oscillograph on a film directly reflect the direction and amplitude as well as the oscillations of

* See Appendix.

## The Beginning of Control Experiments 85

the respective sensations. For example, I can show you the curve of pleasurable sensations, such as those triggered by tickling of the hand (tickling phenomenon), the curve for the taste of sugar on the tongue, this drooping curve caused by the subsequent taste of salt, this curve reflecting the pleasant sensation experienced by two persons when warmly shaking hands, this curve obtained during a kiss, and here, following the kiss of a happy couple, the curve of an unhappy couple in which the female partner does not really seem to appreciate the taste of the kiss, in which you will have no difficulty in detecting the droop of the curve.

I do not want to get involved in a discussion of the practical interest—and a humorist (this is France, after all) would doubtless say the danger—of such a discovery. I mention all this just to give you some idea of Dr. Reich, who is a true scientist with much experience in the laboratory; he has been conducting experiments for the past ten years and consequently is very skilled in applying experimental methods with great precision. I would like to mention a few of the works that Dr. Reich has published: *Character Analysis, Dialectical Materialism and Psychoanalysis,* and *Psychic Contact and Vegetative Current*. He has also written other works dealing with mass psychology and its relationship to psychoanalysis.

My point in making these preliminary remarks stems from being asked to examine and assess—to the extent that this is possible on the basis of fragmentary information taken from manuscripts—a number of studies which have yielded what must at the very least be described as totally surprising results that seem to contradict the most solidly based scientific dogmas. A judgment of this kind necessitates, or, rather, would demand, absolute objectivity. However, as such objectivity is not to be found in this world—nor indeed perhaps anywhere in the universe—it is quite natural that we approach the examination of such a question with certain emotions, which are all the more intense since this question demolishes beliefs that have become a part of ourselves and that we have regarded as definitive ever since we acquired them. Therefore, the atmosphere which surrounds the presentation of such findings is of great importance, not because it is intended to create a favorable bias right from the start, but because the aim is to counteract the established prejudice. In fact, Wilhelm Reich's life and work earn him the right to expect our objectivity, our willingness to reject the same sort of prejudice and overhasty judgment that Descartes condemned, and to demand from us a sympathetic, i.e., a scientific and keenly critical, curiosity.

The central premise of Dr. Reich's work, which was evident even in his first experiments, is the equation of the electrical tension-charge process with the process of vegetative life. With this central idea in mind, he has been conducting research into the origin of life for several years, taking the tension-charge process as his starting point. His work has now led him to conduct experiments on the production of artificial unicellular organisms (*édifices monocellulaires*) in which the phenomena of vegetative life are generated by an exclusively electrochemical process. Naturally, he does not introduce any independent electrical force into the elements which he combines. In his work with colloids, the phenomena of surface tension and molecular motion combined with electricity occur spontaneously, at least in the form of movement, which is regarded as a characteristic of life.

Thus, he has made it a rule to conduct his experiments only with sterile substances; i.e., substances from which all vital signs of life had been eliminated, to the best of our present ability and knowledge. In other words, while relying on radical pasteurization of the substances, he proves the non-existence of radical pasteurization.

Dr. Reich visited me to tell me of his work. I translated some of his writings so that they would be available in France. However, for a year, I had heard nothing more about his detailed research, when suddenly, on January 8 of this year (1937), he sent me a letter and a brief report with two sealed and sterilized ampoules which he asked me to examine under the microscope

## The Beginning of Control Experiments 87

at about 3000×. I did so almost immediately and I shall read you the report that I wrote right after the examination. To begin, however, I should like to read my translation of the detailed report which I received from Dr. Reich. The translation was done in haste and I shall have to ask you to excuse the style, for I did not have time to correct it. I can, however, most definitely vouch for its accuracy. Incidentally, I have brought along the originals of this report and of all the documents which I shall read to you in translation and any of you who can understand German and wish to examine the original texts are very welcome to do so.

On January 8, 1937, I received a preliminary report on the production of lifelike forms in accordance with the tension-charge formula.* I prepared the following summary:

In a letter dated January 8, Dr. Wilhelm Reich, who lives in Oslo, wrote me a preliminary report on the formation of structures which have the characteristics of life based on the tension-charge formula.

The report showed that after many years of research, Dr. Reich, through mere physical and chemical processes, had obtained structures having all the characteristics of life. The report described in a certain amount of detail the experimental procedure; i.e., the constituent elements of the structures as well as the sequence in which the elements were used (this sequence, incidentally, seemed to Dr. Reich to be the most important basis of the experiments).

These reports were accompanied by sealed ampoules, each containing about 5 cc of substance. They were labeled as follows: Bions 6b sterile January 12/37. A short note attached to the ampoules stated: Observe at 2000-3000× under a binocular microscope.

On Tuesday, January 26—i.e., twelve days after the preparation date given on the ampoules—I was able to conduct a series of observations which met the requirements laid down by Dr.

* See pp. 61–3.

Reich. These observations were made possible through the friendly cooperation of Dr. Ronchese and Dr. Saraille, who run a superbly equipped analytical laboratory in Nice. I did, in fact, have a binocular microscope at my disposal that was capable of at least 2500× and up to 3000×. Despite illumination problems, fruitful observation was still possible at the latter magnification. Immediately after an ampoule had been opened, a sample was taken and placed on a sterile specimen slide. A second slide was immediately placed on top of the first and sealed with paraffin to prevent any evaporation, which would have caused the fluid to collect at the periphery.

My observations quickly and accurately revealed what Dr. Reich had described in his report. I was able to observe four principal forms:

1. Rods which at 2500× had apparent dimensions of ½–1 cm. These forms move in two different ways: sometimes they move vigorously in a longitudinal direction, then they suddenly stop and start moving again in an undulating, crawling manner. They give the impression of being fragile bodies that are rather broad and flat, and resemble the shape of certain types of fish. As they swim around, they are seen now from the side, now from on top. They seem to propel themselves through the liquid by undulating their bodies in every direction, just like fish swimming around in an aquarium. These structures possess a nucleus which also moves and vibrates. Furthermore, they undergo division and behave in every respect in a manner similar to living unicellular organisms. I watched several of them divide and produce two similar bodies, passing through the familiar characteristic forms.
2. Unicellular, mushroom-shaped objects with a luminescent, constantly quivering nucleus.
3. Significantly large forms which looked like mycelium with spores on the end of each branch. These forms expand and contract constantly, although very slowly and barely perceptibly, while remaining stationary. Compared with the previous structures, their dimensions are huge, with an apparent length of several centimeters at 3000×.

4. Undivided, anucleate cells which clearly move in a much more mechanical way as if moved by an outside force.

My overall impression is that the preparation contains living microorganisms, although a priori this seems to be at variance with the precisely stated fact that the preparation was obtained by boiling and is thus absolutely sterile.

Nevertheless, it seemed to us that the main experiment which would decide the question whether these forms are living or nonliving would be to propagate cultures of them. Only the progressive elimination of the artificially formed first elements and their replacement by new structures which formed through their own potential would permit us to say whether we are dealing here with dynamic living organisms or whether the dynamism is merely simulated by chemical and electrical processes. I hope that the experiments will be continued until this fundamental question has been settled. At all events, Dr. Reich deserves praise for having come this far.

Roger du Teil<br>Nice, February 3, 1937

I must add to this report that Dr. Rochese, who examined the cell formations with me—and indeed before me—has read Dr. Reich's report very carefully and, although he reserves his interpretation of the observed phenomena, he never for one moment doubted the accuracy of the experiment which produced the bions. This, the opinion of a bacteriologist, deserves to be mentioned here because it reflects favorably on the seriousness and accuracy which all such experiments should possess. And I should also like to add here that M. Deel, a bacteriologist from Cannes, has also read the various reports and examined the cultures, and while he likewise reserves judgment on the interpretation of the results—although his reservations are somewhat different from those of Dr. Rochese—he too had little if any doubt about the appearance of microorganisms in the culture which had been kept for two hours in the sterilizer at 180°C. He even added that the culture which he held looked very much like a "pure" culture, which would exclude the hypothesis that these are so-called air germs. Furthermore, we shall see that in the later experiments particular attention was paid to trying to eliminate these various objections.

On February 8, Dr. Reich wrote again, informing me that he had methodically cultivated the bions and that he would keep me informed of the results because he shared my opinion that the success of these cultures was of extreme importance for the interpretation of his discovery.

On February 16, Dr. Reich wrote again to give me, as requested, all the data needed to start a bion culture. Far from avoiding controls, he invited them. This is a further encouraging sign of his confidence that he has not committed any experimental errors. I will read you the letter now.

Oslo, February 16, 1937

Dear Professor,

I should like to inform you today of some useful and unequivocal results obtained in my bion culture experiments. I agree with you entirely that the question of spontaneous generation cannot be settled by microscopic findings but mainly through successful cultivation of sterile—i.e., cooked—colloidal mixtures. I am happy to be able to report some positive results to you. The matter is simpler than it seemed to me in all the months of difficult experimenting.

I would be extremely grateful if, as promised in your letter, you would repeat and check my experiments at a laboratory in Nice. I am convinced that this can only help me and I am therefore only too happy to comply with your wish to verify my work. The same thing goes for the cultures.

Last week I succeeded in culturing fresh bion preparations on agar nutrient medium and in broth and I found that all four types occurred. It was discovered that fresh heated bion mixtures grow much more slowly and exhibit much less movement in the culture than two-to-five-day-old heated bions. I would recommend that you yourself prepare such a bion mixture according to the instructions in my first letter. You should let the

## The Beginning of Control Experiments

mixture itself heat for about one hour at 160°C in the dry sterilizer and allow the paraffin-sealed ampoules to stand for three to four days. About two droplets should be taken under sterile conditions with a Pasteur pipette from the sterile ampoule and the surface of the agar nutrient medium should be coated with this liquid. After twenty-four to forty-eight hours, either a fine bumpy coating of the same color as the nutrient medium or a dense grayish-white growth occurs. I was unable to determine what causes this difference. The results of the cultivation in broth are much clearer. After only twenty-four hours, the broth liquid becomes very opaque and vigorously moving rods, round cocci, large nucleated cells, and finally amoeboid forms exhibiting internal motion are visible under the microscope.

Yesterday I cooked a broth culture of bions for a quarter of an hour in the sterilizer and even at 250× in the dark field both the motion and shape of the formations were retained. I could not believe my own eyes, but we have repeated the culture experiments so often in different ways that there can no longer be any doubt. I am sending you herewith a culture sample. If you should come to the conclusion that my statements are correct, would you please inform the Academy of them. I will now complete the control experiments, prepare a detailed report, and send it to the Academy with a copy to you.

W.R.

January 19, 1937    *Postscript.* I waited a few days before posting this letter because the control experiments yielded some very unusual results and revealed some remarkable facts. However, it seems that it will take a very long time to complete all the control experiments and I did not want to make you wait unnecessarily. The cooked bion mixtures that have stood for two to three days in the incubator regularly produce very strong growth in broth. Fresh mixtures and freshly heated bions seem to be feeble; i.e., the growth on agar is not as strong as after three to four days. It is also confirmed that bion cultures heated for fifteen minutes continue to produce growth. I have not had one single failure in two weeks of repeated inoculation of broth. I must ask you please to be patient and to wait for the detailed report on these findings and on the control experiments until I have brought the latter to some sort of conclusion. Please write and let me know if you want me to send you sterile bion preparations for cultivation testing or whether you prefer to prepare and cultivate the mixtures yourself. I have so far propagated four strains of various bion mixtures in broth but mainly on agar (strains IV-VII). Strains I-IV were grown on egg-white nutrient medium, but this type of medium was discontinued because of its unreliability and the similarity of the substances. I would therefore be very grateful if you would write and let me know the results of your control tests. Judging by the reliability of the findings, there is nothing to contradict the statement that bions can be cultivated in broth. So far, I have only noticed that the rod form predominates in broth while mainly cocci develop on agar. I do not know what significance this has. It is possible that the nutrient medium exerts some influence on the selection of the various types. I am in the process of finding out what is behind this phenomenon.

On February 22, Dr. Reich wrote to me again, announcing that he was sending a few cultures and some very precise details on how to conduct the experiments.

Oslo, February 22, 1937

Dear Professor and Colleague,

I must bother you with a short postscript to my letter of February 20. The further control experiments have yielded such unusual results that I must stress once more that the experiments involving the sterilization and cultivation of the bions were carried out as follows:

The basic materials which were mentioned in my first report on the composition of bions were heated prior to being mixed and after mixing were placed in glass vessels in a dry sterilizer set at 160°C. The liquids were heated for a half hour to an hour in the sterilizer at 100°C. The cultures were heated for

## The Beginning of Control Experiments

one quarter or one half hour in the same way in the sterilizer set at 160°C. Since it is possible to object that heating for one hour in the dry sterilizer is not sufficient to exclude the possibility of infection from outside or from within by bacterial organisms, we carry out the following experiment:

The dry substances that make up the bions are dry-sterilized for two hours at 180°C. The liquids that are needed to produce the bions are sterilized for one half hour at 120°C in the autoclave and then combined with the dry substances. This mixture is stored under sterile conditions for forty-eight hours before inoculating it into broth. As a second control experiment, we first mix the substances and then autoclave them in the mixed state at 120°C.

The following additional points seem important to me. It is not the aim of these experiments to show that the bions move like living organisms, because that can be clearly established microscopically. The sole purpose of these experiments is to exclude the objection that the cultivation results are due to infection from outside; that is to say, by so-called air germs. Furthermore, in order to test the accuracy of this objection, I am simultaneously conducting various experiments involving inoculating dust from a vacuum cleaner onto nutrient media. So far, the dust cultures have a different macroscopic and microscopic appearance from the bion cultures.

This is all the guidance that I can give you for the moment. I want to assure you that we are proceeding with all due skepticism, but we are also fully prepared to record what we find, and every possible effort is being made to establish that the culturability of the bions is an *unequivocal* result.

With warmest greetings,
Yours,
W.R.

Dr. Reich sent me a telegram on February 25 to tell me that the cultures were successful and had been sent off.

Finally, on February 27, together with a set of instructions,

I received a first shipment of cultures, which was to be followed a few days later by a second. Along with the cultures I also received the following letter and a second report on the culturability of bions, preparation 6 in broth and on agar, together with the positive results of autoclaving.

Oslo, February 27, 1937

Dear Professor and Colleague,

As I told you last week in my telegram, the autoclave experiment was successful. In the enclosure I am sending you some samples of *autoclaved* bion mixtures and a second interim report on the positive result. Also enclosed are instructions on inoculation; a description is given in the report.

May I ask you to send some of the cultures—I will leave it to you which ones—to the Academy in Paris and also to let them have the second copy of the "interim report." I am very anxious to learn whether you were successful in producing and cultivating bions in accordance with the instructions given in my letters. In particular I would like to draw your attention to the strange forms of agar-amoebae which I find astonishing.

With grateful thanks I remain,<br>Yours,<br>W.R.

In addition, you have here the cultures themselves and you can examine them at least macroscopically. Let me tell you that the agar culture, which preferentially assumes this undulating shape, has undergone very strange development during the few days that it has been in my possession and it grows as one observes it. Remembering that the constituent elements were sterilized at 180°C, one is dismayed and rather inclined to doubt the efficiency of the sterilization methods currently employed. I say this without in any way wishing to offend our dear member

---

° See chapter 3.

from the pharmaceutical field nor any of our honored medical members!

This brings me to the end of the objective listing of the facts. Before I hand over the discussion of them to you, however, I would like to add some personal views of my own.

There are two ways of considering these experiments. We can concern ourselves either with the facts or with the interpretation of the facts.

Likewise, the facts themselves can be considered from two different angles. On the one hand, they describe the formation of organized structures from non-organized material, and on the other, they indicate the resistance of these structures to destruction by sterilization. Let us also note something that these two sets of facts have in common: the non-organized elements used in the experiments were sterilized beforehand and resterilization was carried out at each stage in the experiment, so that at each stage it is possible that the non-organized state was restored. Furthermore, this permitted a multiple number of checks to be carried out; with each new cooking, a completely new experiment was performed, establishing whether there was a transition from the organized to the non-organized state. In fact, when I questioned specialists on this matter—and I ask the specialists here today either to confirm or to disprove this information—I was told that, according to the present state of biological knowledge, there is no known microorganism that can withstand a temperature of 180°C. At the very least, Dr. Reich's discovery has shown us that organisms which display all the characteristics of life and which can withstand these temperatures do exist. Now each of us can confirm this discovery with the aid of the bions and the bion cultures which Dr. Reich has sent me.

While we are discussing the facts and their interpretation, we should consider a number of other views relating to the endogenous or exogenous origin of the microorganisms discovered in this way.

As you have already heard, Dr. Reich has two ways of refuting the statement that the bacteria are organisms from the air. First, in the same way that he answers the other objections, he sterilizes the material at 180°C, a temperature which no known bacterium can withstand; second, he takes atmospheric dust from a vacuum cleaner and cultivates it. The cultures obtained in this way have nothing in common with the cultures which Mr. Deel yesterday described as obviously "pure" cultures.

The objection that the culture medium itself might contain bacteria can be countered first by pointing out that the bacteria would not develop just on the inoculated region. The second argument is again the sterilization; and the third response is to point out that such bacteria would be polymorphous and would not present the appearance of a "pure" culture. Having thus presented the facts in a favorable light, we come now to their interpretation, which we expect to see confirmed unconditionally in the near future. Two objections have been made to Dr. Reich's hypothesis that these are truly living organisms, although he himself has not yet stressed this latter point.

The first objection states: What we have here are nothing more than electrical and chemical processes which exhibit movement similar to Brownian movement (this is the chief objection put forward by Dr. Ronchese). Reich's answer is to demonstrate the culturability of the formations.

The second objection, raised by Mr. Deel, is that lecithin is a living substance. In his view, therefore, Dr. Reich has not discovered the missing link in the chain of development from the inorganic to the organic but has merely succeeded in "organizing" a living but still unorganized substance. One can counter this objection by pointing out that lecithin in egg yolk is only a nutrient, while the life force is probably located in the germ cell of the egg. The process of obtaining lecithin, stirring it, dissolving it in ether, purifying the solution with zinc chloride, which forms a practically insoluble double salt from which lecithin is regenerated by hydrogen sulphide, certainly gives the impression of being nothing more than a chemical process in which there are no signs of vital dynamism. And even then, the possibility of life would have to be excluded by the action of sterilization, at least according to the present status of our knowledge. Even if lecithin was something other than the material from which the yolk is composed, it would be killed by this sterilization.

Furthermore, if the objection was made that these are electrochemical processes which completely and perfectly imitate all the manifestations of what we call life, we would have to reply without a doubt: if two triangles prove to be congruent, which is to be regarded as the first and which the second? Which would be the true manifestation of life and which the false? Let me conclude this report, which already contains quite a lot of scientific material for discussion, with a brief excursion into the realm of metaphysics. Let us presuppose that we are in fact able by electrochemical means to construct unicellular organisms which exhibit all the characteristics of life *sensu stricto* as we have conceived of it so far. Perhaps then we should stop regarding life as being manifested by the material phenomena of movement, feeding, and division, and instead see it as a progression from a germ to the organization of differentiated structures corresponding to a type or a genus. And is it not the result of these experiments, which seemed to make us inclined a priori to choose a materialistic solution to the question of life, that we are obliged on the contrary to see life as having an "organizing intention" and to assign it radically to the realm of the spirit?

In practical terms, I believe that our involvement and the interest that we show in this discovery cannot fail to be of great service to the cause of science by enabling Dr. Reich to make his discovery known in France at the Académie des Sciences. It would also help if we acknowledge the accuracy of his experiments and draw his attention to any errors we might find in these experiments or the interpretation of them. Either way we would have made ourselves useful.

Following the discussion, I would therefore like to call on you to appoint several members to assist me in the series of control experiments which I intend to undertake in order to confirm Dr. Reich's results.

Roger du Teil

*Postscript.* At the end of the discussion, the Philosophical Society, in acknowledgment of the obvious interest generated by Dr. Reich's studies, regardless of the interpretation to be put upon them, appointed the following members to assist Prof. Roger du Teil in his task of verifying Dr. Reich's work: Dr. Chartier; Dr. Perisson; Miss Fermand, Associate Professor in Natural Science; and Mr. Claude Saulnier, Pharmaceutical Chemist. In addition, it is the wish of the society that a large French laboratory—in particular, the Lumière laboratory in Lyon—be entrusted as soon as possible with this matter.

Meeting held on Sunday, March 7, 1937

5

# Culturability Experiments Using Earth, Coal, and Soot

## ELIMINATION OF THE OBJECTION THAT PREEXISTENT SPORES ARE PRESENT

The purpose of the control experiments, which have been conducted parallel to the bion experiments for the past two years, was to answer two fundamental questions regarding the interpretation of the results:

1. Are the bions and their cultures perhaps the products of an airborne infection?
2. What is the specific significance of the individual substances from which the bions are created through heating?

The first question can be answered easily by citing certain facts. Bions cannot be generated by an infection from the air during their preparation, because:

1. The motile cocci, rods, and amoeboid structures are present at the moment when the first and second mixture are combined (cf. first preliminary report).
2. Infection from the air is excluded by correct and meticulously careful application of sterilization procedures.
3. The bions retain their motility and culturability, even after long and repeated boiling and autoclaving.

The sterile bions and bion cultures from mixture 6 differ *macroscopically* and *microscopically* from putrefactive bacteria, molds, and other formations that one finds in non-sterile preparations of tissue or grass; the shape, movement, and color of the macroscopic growth are all different.

The second question is not so easily answered: Could it be possible that the bions do not come from the non-organized lifeless substances that are heated together, but from "spores" which might be present in substances such as coal and gelatin and which can withstand heating at 100°C or 120°C? In this case, we would not be dealing with the organization of lifeless material into living forms but merely with the development of hitherto unknown spores.

In the course of the work on heated earth and coal, my original impression that these substances exhibit far more motion or "life" in the heated than in the unheated state was repeatedly confirmed. But the objection that the heating released spores that were already present in the substances cropped up again and again: I not only thought of it myself; it was also raised by bacteriologists. It was claimed that the observed phenomenon was nothing special because the substances simply contained spores which were released by the heating process. Clearly, simple observation of the heating experiment and comparisons with unheated substances would not be enough to refute conclusively the view that preexistent spores were present. I did not doubt for one moment that there are spores from which organisms can develop. However, the assertion that these spores were continually present was contrary to all my observations. How could spores enter coke distilled at temperatures in the thousands of degrees? How could spores appear in blades of grass or in structured striated muscle when they cannot be observed in the unheated substances even at the highest magnifications? And how could they appear *suddenly* when the meat, moss, grass, coal, or earth was heated? Yet, no matter how forcibly this reasoning disproved the metaphysical spore theory, the proof was still theoretical and not based on experiment. I was now faced with the task of inventing a series of experiments that would clarify the matter. The description given on the following pages might lead one to believe that the work proceeded smoothly, without difficulties, and without any worries. In fact, the opposite is the case. No progress was made for weeks and months, and it seemed that the entire experiment had reached an impasse, until a relatively simple idea provided a way out.

## Culturability Experiments Using Earth, Coal, and Soot 101

In the previous heating experiments, the act of sterilization and the swelling of the material had taken place in *one single process*. It now occurred to me that the sterilization—i.e., the killing of any spores present—and the life-generating process of swelling could be separated.

Earth and coal were pulverized separately, placed in Petri dishes, and then put in a dry sterilizer set at 180°C. The standard time for *dry sterilization* at 180°C was two hours, but experiments were also carried out for fractions of this time, or longer, with the same results. Before the two hours were over, sterile test tubes were two-thirds filled with autoclaved 0.1N KCl. The test tubes were tightly sealed with hydrophobic cotton wool. For even greater certainty, the already sterilized potassium chloride was autoclaved once more for half an hour at 120°C. Next, a glass or metal spatula was slowly heated to red heat several times in a gas flame. Then the dry sterilizer was opened and a test tube containing potassium chloride was unsealed. Pulverized coal or pulverized earth was then taken on the tip of the spatula and transferred as quickly as possible to the test tube containing KCl. This was then immediately resealed and placed in the sterilizer. (One could also dry-sterilize the earth or coal in the test tubes and then add potassium chloride.) The test tubes were placed on a slant in order to present a larger surface area of liquid. Experience has shown that this facilitates the disintegration of the pulverized earth and coal. The KCl-coal or KCl-earth mixture was then heated. It was not important how long the substances were heated, only that *both the earth and the coal solution assumed a turbid, dense, colloidal character as a result of the heating process*. In some cases this happened after only a half hour, while in others it took an hour or an hour and a half of heating, depending on the quantity or particle size of the earth or coal.

The most important aspect of the entire procedure was to achieve, if only for a short while, a stable suspension of coal or earth particles in the liquid. The experiment does not give a positive result if:

1. The test tube is taken from the dry sterilizer and the coal or earth is at the bottom of the tube and the supernatant liquid is clear.
2. Suspended particles of coal or earth slowly or quickly settle out when the test tube is taken from the dry sterilizer.

A positive result is achieved only if the particles remain in suspension for at least five minutes. Even in preparations that have been thoroughly heated for a long time, the particles tend to settle out sooner or later.

With a dry-sterilized Pasteur pipette, a few drops are taken from the cooked or uncooked coal or earth suspension and are examined microscopically and electrically. Two things are discovered: Upon *electrical examination,* the coal as well as earth particles move toward the cathode when a 2 mA current is passed through the preparation (see elsewhere for a description of the technique). If the current is reversed, the direction of motion also promptly changes. Similarly, the motion stops immediately and completely when the current is interrupted.

If the current is passed through for a long time, the particles acquire a *negative* electrical charge, as did, for example, staphylococci.

The *microscopic examination* is carried out in two ways: First, the objects are observed in a dark field using a 10× objective and an approximately 16× compensating eyepiece with angled binocular focusing tubes. When the dark field is correctly set and the densest middle layer of the preparation observed, vigorously moving individual vesicles and clusters of vesicles are seen. When the magnification is increased to at least 2000×, we see several crystals of coal or earth lying still. Most of the coal and earth crystals are green in color and have started to break down into vesicles, particularly at the edges. Tubular structures can be observed moving, expanding, contracting, bending. The vesicularly structured coal and earth formations crawl around in the field of vision with jerky movements. Rods and cocci pass across the field, moving rapidly with a quivering motion. The

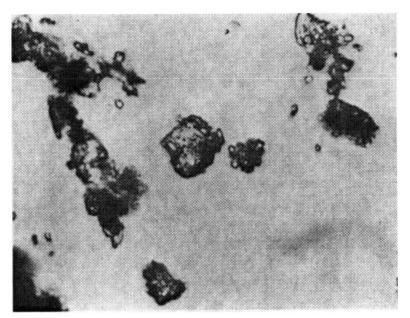

Figure 48. Garden earth, dry sterilized at 180°C. 1000×

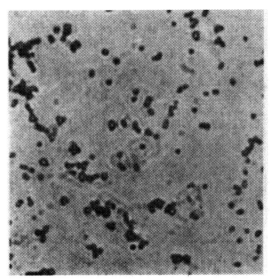

Figure 49. Culture of this garden earth I db 16. 1000×. Gram-stained

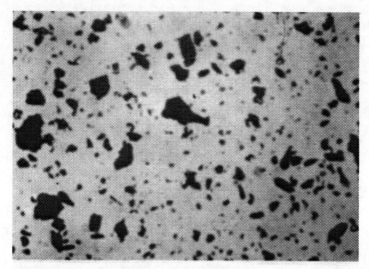

Figure 50. Coal dry-sterilized at 180°C. 1000×

Figure 51. The same coal vesicularly disintegrated after eight weeks in potassium chloride. Time-lapse photograph from film preparations 1, 2, and 3

movement is typical of unicellular organisms. If the suspension was incomplete, the same formations would move either less vigorously or not at all. Obviously, long-term suspension of the particles in the liquid, positive electrical charge, and motility all go together. If the preparation exhibits these three basic parameters, inoculations are made from the cooking colloidal solution into beef broth. After twenty-four hours, in many cases the earth as well as the coal inoculations produce a dense milky turbidity in the broth. Microscopic examination reveals that mainly rod-shaped, rapidly or sluggish-moving objects are present. The coal- or earth-amoeboid formations which were taken from the heating preparation on the previous day are no longer present. Occasionally, one sees unstructured crystals from the inoculation. Next, under sterile conditions, a drop is taken from the broth culture and coated on an agar nutrient medium. A yellowish granular growth forms after twenty-four hours and often sooner. As a control of this growth, two other cultures are prepared:

1. One agar nutrient medium with unsterilized, pulverized coal, and one with unsterilized, pulverized earth.
2. One agar nutrient medium with pulverized earth dry-sterilized at 180°C, one with pulverized coal dry-sterilized at the same temperature.

The first control experiment produces an irregularly shaped growth of various colors; the second control produces no growth at all. The sterilized agar growth of the coal or earth mixture which was first sterilized and then cooked differs microscopically in every respect from the non-sterile agar growth. The more closely the following empirically derived treatment methods are adhered to, the more possible it is to guarantee the culturability of sterile earth and coal immediately following heating:

1. The coal or earth crystals must first be finely ground by mechanical means.
2. The evaporating KCl solution must be replenished under sterile conditions.

3. The liquid must be uniformly colloidally opaque; the particles should not settle out completely.
4. It should be ascertained by microscopic examination that mainly crawling structures are present.
5. It is best to inoculate immediately from the cooking solution.
6. The first generation should be inoculated into broth and the second generation onto agar. For better results, inoculate first onto egg medium and then onto agar.

Let me now summarize the errors which may prevent the growth of cultures of earth or coal:

1. If the coal or earth is sterilized too long without being heated for a suitably long period of time, the vesicular disintegration of the crystals remains inadequate.
2. If no attention is paid to whether the particles settle out when taken from the sterilizer, then the cultures fail easily.
3. If the test tube is taken from the sterilizer without due care or, even worse, if it is shaken, the substance lying on the bottom is disturbed and only gives the impression of being in suspension. The test tubes should therefore be carefully removed and should not be shaken.
4. Unless one examines the preparation immediately after heating, one cannot know whether the structures show significant movement.

The coal treated in this way, as well as untreated coal and moss, did produce a few cultures in broth. Though, on the whole, the results of these cultures were uncertain, the culture test was one of the most decisive elements in the entire series of studies. If a culture was unsuccessful, there were two possible explanations:

Either the substances, which had been completely dried out by the dry sterilization, were not heated long enough to achieve the necessary breakdown and swelling of the particles, or the nutrient substance was not complex enough to fuel and sustain incipient life. On the other hand, the microscopic findings were always the same, provided I carried out heating until colloidal turbidity was achieved. If the preparations d/b are allowed to stand in sealed ampoules for two to three months, then it is often an easy matter to produce cultures. So far, it has not proved possible to say why the culture succeeds on one occasion but fails on another.

Continuing experience with the coal and earth experiments showed that the reliability of culturing increases when *higher temperatures and smaller particles* are used.

## THE INCANDESCENT COAL EXPERIMENT

On May 21, 1937, it occurred to me that there was a very simple if radical way of checking the metaphysical spore theory. One could devise an experiment with coal which would entirely eliminate any possibility that spores are present. The experiment which I carried out is very simple. Provided that one uses a binocular microscope capable of 2500–3000× with an immersion lens system, the observed phenomena are unambiguous. Let me mention in advance that this experiment was repeated ten times in immediate succession, always with the same result.

Some of the coal dust which remained in the dry sterilizer at 180°C was placed on a thin metal spatula. It was then heated to incandescence in a benzene gas flame—i.e., to about 1500°C—until all the coal dust was incandescent without becoming completely ash. In the meantime, my bacteriology assistant prepared a mixture of beef broth which she mixed in equal parts with autoclaved 0.1 N potassium chloride. I proceeded from the assumption that not only nutrient substance but also a swelling substance should be present. In this experiment the incandescent coal was immediately introduced into the broth-potassium chloride solution. To our surprise, in all the experiments the solution immediately became colloidally opaque in a way that had never been achieved so well after simple heating.

Another instructive and surprising phenomenon was observed. The particles of coal initially gave the solution a blackish color, but after ten to twenty minutes this black coloration was replaced by a cloudy gray turbidity. This solution was examined under the microscope at 3000× immediately and then after twenty-four, forty-eight, and seventy-two hours. Very vigorous life forms were evident *immediately* after the preparation had been made. A dry sample of coal dust heated to incandescence was placed on the slide. Even at 200×, it was possible, in the dark field, to see the most delicate vesicles into which the crystals of coal disintegrated. When potassium chloride was added, motility set in. Some of the vesicles, which resembled spores, moved around. The large particles of coal greedily soaked up liquid. After one to two minutes, the large particles of coal, whose borders had become completely vesiculate, were observed drawing individual vesicles toward themselves from the fluid. The movement of the individual vesicles toward the large coal particles speeded up as the distance diminished. Finally, the individual vesicles darted toward the larger coal particles and adhered to their edges. Without a doubt, these were electromagnetic phenomena.

At 3000×, it was not possible to distinguish coal particles heated to incandescence from living formations. The particles crawled around in the field of vision, with significantly more movement in the middle and upper layers of the preparation than in the bottom layer. After twenty-four hours, the potassium chloride solution was full of moving cocci and rods, found to have a *positive electrical* charge. The bluish-black color of the structures was an immediate indication of their origin. The cocci were all sizes. The margins of the larger formations quivered strongly even after the preparation had first been cooled. This movement was therefore not a heat phenomenon. The first preparation produced dense turbidity in broth after twenty-four hours. No turbidity occurred in the other preparations, and inoculation resulted in only a few cultures. To provide a solid base for my further studies, I first had to find a provisional explanation for the observed phenomenon.

In coke crystals, which have already undergone high-temperature distillation, the individual particles cohere strongly to one another. After heating—particularly after heating to incandescence—this cohesion breaks down and electrical energy is released. (Otherwise, the individual particles could not be electrically charged, and the large specimens are not.) When potassium chloride is added, the particles into which the calcined coal disintegrates are seen at 2000× to take up large quantities of liquid. These particles change into vesicles indistinguishable in shape from spores. To my mind, this experiment utterly refutes the spore theory as it now exists, for no one can prove that spores can survive the temperatures generated in calcination. To this must be added my other findings: all the substances examined assumed vesicular structure when boiled, autoclaved, heated to incandescence, or made to swell. And, as I have already said, the vesicles are identical in appearance to spores. I do not deny the existence of spores within the meaning of the old theory; I merely claim that the vesicular character and the disintegration of swollen, boiled, or calcined unorganized matter into vesicles is the central point from which to proceed in further examining the *formation* of viable spores. *Thus, the spores, too, must originate from matter as a result of swelling.* It seems to me it is absolutely essential to make this assumption.

In fact, the accuracy of this assumption can be proved by cinematographic means. Fine coal dust is first dry-sterilized for two hours at 180°C. A few granules are then placed in a hollow slide and autoclaved KCl is added. A cover glass is immediately placed over the preparation and sealed with paraffin. The microscope is then set at 300–600× with dark-field illumination. Next, about 50 cm of film are exposed at normal speed. Then the time-lapse mechanism is set at one frame every two or five hours. For time-lapse photography it is best to select a *flat* crystal which is as free of vesicles as possible and has a sharply defined border. The time-lapse camera operates without interruption for 6 to 10 weeks. Daily observation merely with the naked eye reveals the progressive vesicular disintegration of the crystals. As time goes by, more and more motile vesicles appear in the field of vision.

When the film is then run through, one first sees a dark spot, which slowly or rapidly gets lighter, depending on the time-lapse setting.

## CULTURES OF SOOT HEATED TO INCANDESCENCE. THE BIOLOGICAL INTERPRETATION OF BROWNIAN MOVEMENT

A typical objection to the movement observed in the bion cultures being organic in nature is that this is a physical phenomenon known as Brownian movement. Brown himself is supposed to have regarded the motion, which he observed in particles of India ink, as a sign of life. Physicists explain the phenomenon in terms of the physical action of molecular movement. Whether or not Brownian movement was involved could only be settled experimentally: *Are the particles which create the impression of organic movement in the solution of India ink culturable or not?*

Initial tests to cultivate autoclaved India-ink solution directly on agar were not successful. Therefore, after several failures, I carried out the following experiment:

Soot was first dry-sterilized for three hours at 180°C; beef broth was allowed to stand for twenty-four hours in test tubes in the incubator in order to guarantee the sterility of the broth. Several blood agar and agar media were inoculated with soot dry-sterilized at 180°C as controls. If my theory was correct, this mixture should not produce any growth. The results were as expected: dry-sterilized soot does not produce any growth on agar. The soot was next heated to red heat for two to three minutes over a benzene gas flame and then immediately placed in one of the broth test tubes mentioned earlier. In the course of about ten minutes, the blackish dense turbidity turned gray, just as in the case of the incandescent coal. After twenty-four hours in the incubator, the broth solution, which exhibited a densely yellowish-white turbidity, was inoculated onto egg nutrient medium. Microscopic examination of the broth-soot solution revealed vigorously moving bions with the characteristics of coal bions. Over a period of twenty-four to forty-eight hours, a growth composed of individual hillocks of uniform color occurred. The broth had promoted swelling of the soot particles and the egg medium supplied the particles with the various nutrients. Then inoculations were made from the coating on the egg medium onto blood agar and agar. After only twelve hours, a bluish-white, creamy soft growth appeared. At 3000×, the microscope revealed formations different from those in the broth; they still clearly showed signs of having developed from bluish-black soot particles, but their movements were vigorous and smooth. Soot particles can thus be cultured without losing any of the conditions necessary for life. The movement which Brown had seen was the movement of soot (lampblack) bions from which life can develop. I believe I am fully justified in regarding the above experiment as a suitable refutation of the objection that the motion observed is physical in nature.

In the same way, I was able to culture ash from a central heating furnace and from charred wood. Under the microscope, the cultured formations had the same characteristics as the soot cultures.

Of eighteen consecutive experiments in which soot was heated to incandescence, only five failed, for reasons I was not able to identify. It was later found that inoculation onto egg nutrient medium may produce only one or two small pinhead-sized round elevations *which must be spread with a heat-sterilized platinum wire in order to obtain a dense growth*. In the case of doubtful egg growths, I let the nutrient medium stand for eight to fourteen days, stored in a horizontal position, at room temperature; eventually, the hummock growth occurs.

The results of the incandescence experiment are surprising only if one assumes that the living structures observed following swelling were already present in the form of spores, because it is impossible to explain how spores could survive such temperatures. Yet this view is incorrect. The vesicular elements—let us not hesitate to call them *bions*—from which the bacterialike

Figure 52. Culture of soot heated to incandescence. Egg nutrient medium. Fourteen days old

Figure 53. Culture of soot heated to incandescence. Blood agar medium, fresh

Figure 54. Non-sterile uncovered agar, two months old

Figure 55. Non-sterile uncovered agar, six weeks old

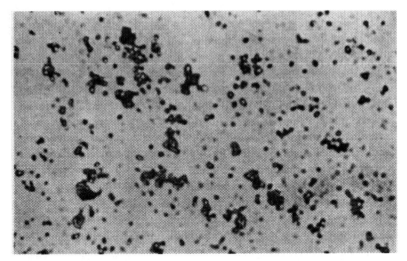

Figure 56. Bions from soot heated to incandescence in broth and potassium chloride

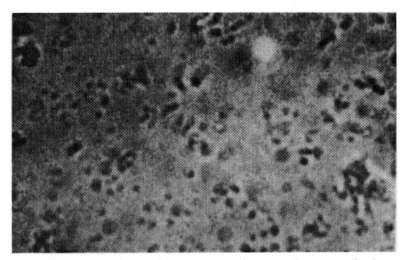

Figure 57. Culture of bions from soot heated to incandescence. Blood agar medium

structures form were not preexistent at all, but formed only after the substances had disintegrated following incandescence and subsequent swelling. I could not help but think that all known types of spores must have formed when the earth was still glowing hot. They must have developed when the previously glowing matter came into contact with water and substances capable of producing swelling. I ground some pebbles to fine dust, held the dust in a benzene gas flame until red hot, and then added potassium chloride to the incandescent dust. Again I saw vesicular disintegration and motile vesicles, which were only very short-lived. So far, it has not been possible to produce cultures.

In the debate on this subject, the view that life is something completely separate from non-life has contrasted with the view that "in the final analysis everything is living." One cannot do much with such generalizations. For the time being, we must say that probably all unorganized matter has the ability to produce life, depending on its composition and environment. *Life can burst forth anywhere for a short or long period of time, given the right conditions.* Sometimes one can observe lifelike motion that lasts only a few seconds. Sometimes the same motion takes weeks or months to die out. Whether or not incipient life survives probably depends on two basic conditions:

1. The chemical composition of the matter.
2. The environment—i.e., the culture medium—in which the spark of life develops.

# 6

## *Control Tests and Instructions for Verifying the Bion Experiments (Summary)*

For accurate verification of the experiments and procedures described here, it is essential not only to follow the rules of complete sterilization but, in addition, to create all the conditions necessary to permit unorganized matter to become organized. The following list of requirements is far from exhaustive:

1. It is essential that the microscope be binocular, with inclined tubes, and capable of providing a magnification of 3000×. Inclined binocular focusing tubes are important because this arrangement not only gives 50 percent greater magnification than the straight single-tube type but also provides a more three-dimensional image.
2. In addition to time-lapse cinematography of the development process, it is essential to observe one point in the preparation for several hours at 2500 to 3000×. In this way, the vesicular changes in the tissues and crystals, as well as the transformation in the shapes, can be observed directly.
3. In order to assess the life-promoting effect of cooking at high temperatures or heating to incandescence, one must continuously compare the unheated and chemically untreated substances with the cooked or incandescent substances. This is the only way to establish that the motile microbial formations in the heated substances do not develop from spores but are created by the disintegrating effect of heat and swelling.
4. The same substance must be examined in the *nonmotile* as well as the motile state. This would disprove the assertion that the movements observed by Brown are unchanging physical

molecular motions because they can exist under one set of conditions but not under the other.
5. The organic character of all movement observed under the microscope must be verified by carrying out a culture test of the relevant substance. If the test fails, it does not necessarily mean that the substance in question is incapable of life: first, the substance may not, in fact, be organizable; second, it may be organizable but the necessary conditions may not yet have been fully identified.
6. It is essential to eliminate judgments based on one single set of results; e.g., those obtained from microscopic observation. A correct interpretation can only be arrived at by simultaneously considering the results of *all* the examinations; i.e., microscopic examination, electrical tests, color reactions, culture experiments, etc., that make cultivation possible. *To promote organization, it is first necessary to follow all the rules of sterilization and to kill off all existing life.*
7. Serial studies with individual as well as with combined substances provide a further indication of the validity of the experiments. Unsterile cultures may be similarly prepared, but we will find that it will take days, even weeks (e.g., if agar nutrient media are exposed) to form a clear culture.

Naturally, the above list far from exhausts the wealth of necessary measures still to be tried.

It is of course important to remember that there are two sources of error which must be eliminated in these experiments. First, errors may be caused by insufficient sterilization, leading one to mistake formations in the culture or in the bion mixture as organized organisms, whereas in fact they result from an infection. But the second error is just as dangerous. One may overlook or fail to reproduce the conditions necessary for lifeless matter to become organized, or perhaps no attempt is made to find out exactly what conditions are needed.

Any discussion of the bion experiments not based on a thorough understanding of the conditions would be fruitless and should be avoided. On the other hand, a great deal of scientific

## Control Tests for Verifying the Bion Experiments 117

information can be expected from controls carried out on the basis of these experiments, thus temporarily dispensing with the usual experimental concepts and thereby helping to clarify many still unanswered questions.

*Summary.* Let me now summarize the most important control experiments that I carried out parallel to the bion experiments of preparation 6:

1. The individual substances from which preparation 6 is made are sterilized in a manner depending on the type of substance and individually inoculated into broth. When autoclaved, the individual substances do not produce any growth.
2. After each new substance is added, a sample is taken of the sterile solution and inoculated into broth. Substance mixtures a + b, a + b + c, a + b + c + d, etc., should not produce any growth. The lack of growth in this control experiment is positive proof of the sterility of the bion mixture. When autoclaved, the *total* mixture must produce growth after three to eight days, if all conditions are met. *It must have a strong electrical charge.*
3. When heated for half an hour or one whole hour in KCl, coal and earth frequently produce a growth in broth. Coal and earth dry-sterilized for one hour at 100°C and inoculated onto agar should not produce any growth. This proves that the swelling of the substance is an essential precondition for vegetative life to develop.
4. In order to compare the structures obtained by sterilization and high temperatures with air germs, the agar is left uncovered. The formations are totally different both microscopically and macroscopically. It takes days and often weeks before a strong growth appears.
5. All the substances used to produce bions are inoculated, completely unsterilized, onto agar and into broth. These agar growths are completely different from the formations that develop in the total mixture after thorough sterilization.
6. After a few days, moss and grass soaked in water are inoculated, completely unsterilized, onto agar. Similarly, unsterilized

water from the tap is inoculated. In every case, the growths differ from all types of sterile bions.

## POSSIBLE FUTURE STUDIES

In this first detailed report, I will content myself with describing verified experimental results. However, in the course of my work, many phenomena appeared which either could not be classified or indeed were confusing. Also, I noted other manifestations which opened up new fields of study.

If every substance breaks down into motile vesicles when exposed to swelling or disintegrative processes, it is logical to assume that the *foodstuffs* which we ingest in cooked form are subject during digestion to an as yet unknown process that is intimately related to the bions. Experiments carried out with *pepsin* on various foodstuffs have in fact produced the same results obtained by heating and swelling: the pepsin breaks down the substances, and their structure is destroyed and replaced by a *vesicular structure*. Detached individual structures and clusters of vesicles are seen moving around. Fresh *lymph* examined in the *dark field* at a sufficiently high magnification resembles a bacterial culture, judging by the profusion of vital movement. These two observations strongly suggest that, *in the animal organism, food is taken up into the blood and lymph circulation in the form of electrically charged vesicular units and clusters of vesicles*. I would stress that this is a very plausible assumption, although it still has not been proved or disproved by experiment. If such an experiment succeeded, it would be of far-reaching importance, indicating that the body obtains its energy in the form of *bions* by way of the digestive process, in the same way that our org-animalcule ingests the individual vesicles. It is conceivable that the cells of the organism incorporate these vesicles as carriers of electrical energy. It would be pointless at this stage to say any more about this or to put forward any further theories.

*Highly sterile bions and bion cultures have a visible effect on various types of bacteria and cells,* and this opens up another perspective for further studies. It has not yet been established whether the biological activity or the type of electrical charge is the effective factor. However, from the phenomena observed so far, it is evident that all manner of combinations of tests can be performed to identify the effect.

Similarly, the idea suggests itself that certain types of bacteria enter the organism in two different ways: either through infection—i.e., the invasion of the body from outside by an organism which propagates itself and exerts a destructive effect in the body—or from still largely unexplained *inner self-destruction of the organism,* in the course of which protozoal forms develop from the tissues. In the case of tuberculosis, the endogenous nature of the Koch bacillus has long been asserted.

The earth-potassium chloride experiments provide new insight into the well-known process of potash fertilization of the soil. It is probable that the potash greatly promotes swelling and the release of electrical energy in the soil so that life energy is released. It is possible that the resulting soil bions have a fertilizing effect. This assumption is difficult to dismiss.

I want to add briefly that I was able to produce culturable bionlike structures from human blood.\* I will give a separate report on this soon. Together with the knowledge of sexuality obtained in my clinical work on the autonomic functions of the human organism, the methods of bion research proved particularly fruitful for an understanding of cancer. Appropriate experiments were initiated eighteen months ago and are yielding promising results.

The work suffered as a whole because *too many* questions were raised simultaneously and because I did not possess adequate means to pursue them. I can only hope that it will not be too difficult to set up a sufficiently large and well equipped laboratory, together with a scientific staff, to tackle the new and complex questions. In any event, my work did not suffer from a lack of problems and solutions but from the difficulty in limiting the flow of new findings.

\* See Appendix.

| Substance | Type of sterilization | Type of swelling | Electrical reaction | Movement | | Culture in | | Electrical reaction of culture | Animal experiment | Remarks |
|---|---|---|---|---|---|---|---|---|---|---|
| | | | | Immediately | 2-6 weeks | 2-6 days | 3-10 weeks | | | |
| Ordinary earth | a | O | O | H$_2$O O | + | O | O | O | ? | Non-sterile mixed cultures |
| " | b | KCl b | ++ | ++ | +++ | ++ | ++ | ++ | ? | |
| Cultured earth | b | KCl c | ++ | ++ | +++ | ++ | ? | ++ | ? | |
| Ordinary earth Cultured earth | c | KCl c | +++ | +++ | +++ | + | ? | ++ | – | |
| " | d | O | O | H$_2$O O | + | O | O | O | ? | |
| " | d | KCl c | ++ | +++ | ++ | ++ | ++ | ++ | – | |
| Powdered coke | a | O | O | H$_2$O O | + | O | O | O | ? | Non-sterile cultures |
| " | b | KCl b | ++ | + | ++ | + | ? | + | ? | |
| " | c | KCl c | ++ | ++ | ++ | + | + | ++ | – | |
| " | d | O | O | O | O | O | O | O | ? | |
| " | d | KCl c | ++ | + | ++ | ++ | ++ | ++ | – | |
| " | e | Bouillon + KCl | +++ | +++ | ++ | ++ | ? | +++ | – | |
| Soot | a | O | + | + | O +? | O | O | O | ? | Non-sterile mixed cultures |
| " | c | KCl c | +++ | +++ | +++ | ++ | ? | ++ | ? | |
| " | d | O | O | O | O | O | O | O | ? | |
| " | e | KCl + Bouillon | +++ | +++ | +++ | +++ | ? | +++ | – | |

Table showing the culturability of ordinary earth, cultivated earth, coke, and soot. The designations have the following meanings:

| | |
|---|---|
| *a* | Non-sterile |
| *b* | Boiled for ½ hour at 100°C |
| *c* | Autoclaved for ½ to 1 hour at 120°C |
| *d* | Dry-sterilized for 1 to 2 hours at 180°C |
| *e* | Heated to incandescence for ½ to 1 minute in benzene gas flame |
| *bb* | Double boiled |
| *cc* | Double autoclaved |
| + | Weak movement, little movement, occasionally positive reaction |
| ++ | Strong movement, frequently positive, clearly positive reaction |
| +++ | Very vigorous, mainly positive result |
| – | Animal experiment without any visible effect |

## TABLE SHOWING THE ELECTRICAL CHARGE OF VARIOUS BIONS AND THEIR CULTURES

| Type of bion | Charge | Culture | Charge | Remarks |
|---|---|---|---|---|
| Mixture prep. 6 bions | negative | yes | negative | Exception: 6 cI. = electr. + |
| Earth bions | positive | yes | positive | |
| Coke bions | positive | yes | positive | |
| Soot bions | positive | yes | positive gram + | |
| Muscle tissue | negative | so far, negative | ———— | |
| Liver bions | negative | yes | negative | ⎫ Culture findings still need further checking and interpretation |
| Lung bions | negative | yes | negative | ⎬ |
| Yolk bions | negative | no | | ⎭ |
| Moss bions | negative | possible | negative | |
| Grass bions | negative | ? | ? | |
| Cow's milk | negative | possible after autoclaving at 120°C | negative | |

PART TWO

# The Dialectical-Materialistic Interpretation

# 7

## *The Problem of the Mechano-electrical Leap*

There is no possibility that today, or even in the near future, we will be able to grasp all the conditions which lead from mechanical filling to electrical charge. The second half of the four-beat life formula, electrical discharge → mechanical relaxation, is easier to comprehend if only because of the purely physical principles involved and because it contains the first part of the biological function. If an electrical charge is present, this gives rise to the function of discharge; if the charge is associated with swelling, then the discharge must go together with relaxation. We are already able to answer some of the questions by drawing on existing scientific knowledge; but the rest will have to wait for an explanation at some future date. What we know is that there are substances capable of swelling; where they exist, life *may* form. We know also, from biochemistry, that *it is the electrical charge of the particles* which keeps the organic colloid in suspension.

### CHEMICAL PRECONDITIONS OF THE TENSION-CHARGE PROCESS

Out of eighty inorganic elements, relatively few are life-specific. According to Hartmann, they can be divided into three main groups:

1. C      Carbon
   N      Nitrogen
   O      Oxygen
   H      Hydrogen

2. P          Phosphorus
   K          Potassium
   Ca        Calcium
   S          Sulphur
   Fe        Iron
   Mg       Magnesium
   Also, in animals: Na (sodium) and Cl (chlorine)
3. $H_2O$     Water

We must now assume that each of these groups of substances is related in some specific way to the tension-charge process and that this relationship determines that these and no other substances are needed to establish the life function. These substances must inherently contain the function of tension and of charge in a particular way.

Let us summarize the life function once more from another angle. All life is fundamentally governed by:

charge      / discharge
swelling    / relaxation
anabolism  / catabolism
tension     / relaxation
assimilation / dissimilation
growth      / dying (shrinking, relaxation, discharge)

We will now try to find the functions associated with the individual substances. In so doing, we will rely entirely on the knowledge yielded hitherto by biochemistry.

Of the substances that control the function of living organisms, *carbon* and *nitrogen* are particularly significant, because they are antithetic. Carbon has the property of forming complex compounds with itself and other organic substances to produce, in particular, proteins. *It thus builds up and binds energy.* Nitrogen, on the other hand, has the property of forming very unstable compounds and consequently of releasing energy. It is therefore justifiable to assume that the fundamental functions of living organisms—namely, tension and relaxation or accumulation and discharge of energy—have their basis in these antithetical properties of carbon and nitrogen.

## ELECTRICAL CHARGE AS A CHARACTERISTIC OF COLLOIDS

If a parchment membrane filled with an aqueous solution of starch and sodium chloride is suspended in a vessel containing distilled water, sodium chloride is found after a short while in the water of the outer vessel. The salt has thus permeated through the membrane. On the other hand, the dissolved starch does not get through. On the basis of this different response to membranes, a distinction is made in colloid chemistry between *crystalloids,* which can diffuse easily through animal membranes (e.g., salt), and *colloids,* which cannot pass through a membrane (e.g., starch). Colloids are the most important constituents of living matter. They can be divided into two different types: lyophobic and lyophilic. Lyophobic colloids are those which, when desiccated under constant temperature, do not leave behind any dry residue and which, when combined with the solvent, do not by themselves produce a colloidal solution again (see, for example, the colloidal metal solutions). Protein, on the other hand, forms a colloid that, when dried and combined with water, produces another colloidal solution: lyophilic colloid.

This property of lyophilic colloids is of decisive importance for understanding the "germ theory." If protein, the principal component of living substance, can dry up and then produce an organic colloid again when it is combined with a specific fluid, it is highly probable that unicellular organisms can dry up, become lifeless, and then be restored to a living, functioning colloid when recombined with liquid; i.e., they can reemerge from the "germ state."

Another property of a colloid that distinguishes it, for example, from a solution of sodium chloride is its lack of homogeneity. In the salt solution the molecules are uniformly distributed; in the colloid the density of the molecular arrangement varies. This is manifested by what is known as the Tyndall effect, wherein if a parallel beam of light is passed through a solution, a significant difference is observed between crystalloid and colloidal solutions. If the beam passes through a protein solution, the path of the beam shows up as turbid and opaque; i.e., the incident light is scattered. This phenomenon is not exhibited by crystalloid solutions.

Biochemistry has produced a number of explanations of the *electrical behavior of colloidal solutions* which are extremely important for understanding our experiments. If an electrical potential gradient is generated in a colloidal solution, the particles of the colloid will migrate; i.e., they themselves will carry an electrical charge. The sign of the charge depends on the chemical nature of the colloid and on the presence of an electrolyte. Biochemists are still not certain what causes the electrical charge of very fine colloidal particles. It is assumed that the charge is produced by the buildup of ions from the electrolyte at the surface of the particles. Thus, a colloid that adsorbs hydrogen ions must have a *positive* charge. Our experiments would seem to indicate that the particles themselves are electrically charged as vesicles. It is also conceivable that the atoms are vesicular in character. This view is supported by most recent research into the structure of atoms, which are shown to be made up of a nucleus and encircling electrons. One could derive far-reaching conclusions from this for the formation of vesicles.

Another biochemical finding is extremely important: The stability of colloidal solutions depends on the particles being electrically charged. *If the smallest particles of a colloidal solution are not electrically charged, then the colloidal solution is no longer stable; i.e., the particles are no longer held in suspension and they sink to the bottom.* The explanation for this is that there are two opposing forces in play. Forces of attraction tend to combine the particles; i.e., to precipitate them out. On the other hand, electrical charges of the same sign give rise to repelling electrical forces. Thus, if the particles are electrically charged, the colloidal particles cannot combine, the protein cannot precipitate out, and the solution is stable.

This explanation seems very plausible to me. In fact, in experiments on bions from heated preparations, the amoeboid motility and the dancing locomotion of the bions and the clusters of vesicles continue only while the particles are moved by electrical forces. After a few weeks, the sediment of the preparation becomes thicker, the solution becomes clearer, and microscopic examination shows that the cataphoretic, as well as the general, motility has disappeared.

This forces us to a further conclusion: the quivering, dancing motion and change of position of the bions is connected with the force effects of electrical fields.

## ELECTRICAL CHARGE AS A PREREQUISITE FOR VESICULAR MOVEMENT

Inorganic substances such as metals, ores, etc., are "immobile" in the biological sense; they do not contract or expand or move from place to place. The hypothetically assumed molecular and atomic "movements" do not constitute any of the three stated fundamental characteristics of motion in organized living matter. Frequent attempts have been made in model experiments to imitate organic motion—for example, the plasmatic flow of the amoeba. These model experiments were based on the assumption that organic motion was controlled by changes in the surface tension, but this explanation answers only part of the question. It is still not understood how the spontaneous impulse to move comes from "within." The change in surface tension is certainly a fundamental phenomenon of movement, but it cannot be regarded as the cause. At this point, a vitalist would introduce a metaphysical principle to close this gap in our knowledge. Bearing in mind the observations and experiments that have been described up to this point, let us try to arrive at an explanation of this internal impulse.

According to our present knowledge, electrical energy in

iron, for example, is bound by a certain arrangement of the molecules. By applying an external electrical force to the piece of iron, one can change the position of the molecules so that the iron becomes magnetized. It is important to remember here that the electrical energy is *uniformly* distributed over the molecules and atoms in the same way as in a crystalloid solution. As already stated, organized matter differs from non-organized matter primarily in that the former is *not homogeneous;* i.e., *of non-uniform density.* Organic colloids are not homogeneous, as is demonstrated by their ability to refract light. This difference between organized and unorganized matter must be of fundamental importance as regards motility or immotility.

If we examine the vesicles of various substances in different states, four basic states pertaining to movement can be discerned.

1. The vesicles *do not move* and therefore do *not* change their position relative to each other (vesicles in gelatin, dust, unheated soot in water, blood charcoal in sodium chloride, etc.).
2. The vesicles *move back and forth on the spot* in an almost rhythmic manner. They do not move from place to place, nor do they influence each other (e.g., in unboiled milk).
3. Movement from place to place, e.g., in the case of streptococci and staphylococci. In this type of motion, the various vesicles exert a force on each other. If an electric current is applied to the staphylococci long enough to give them a negative charge and render them immotile, then with time they also lose their cataphoretic ability. Thus, motion and electrical charge must in some way be related.
4. *Movement within the organic system;* i.e., relative displacement of the parts of the organism. This mainly takes the form of expansion and contraction.

The lack of motion in vesicles and "movement on the spot" are a separate problem and will not concern us here. The experiments yield the following plausible explanation for the types of movement in 3 and 4. Even at maximum magnification in the dark field, unheated and unswollen crystals of earth do not exhibit any

### The Problem of the Mechano-electrical Leap

Forms of visible motion

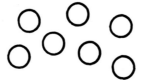

(1) Vesicles lying still

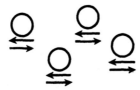

(2) Vesicles moving in place

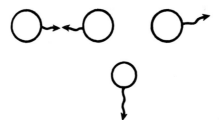

(3) Vesicles moving from place to place

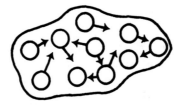

(4) Amoeboidally motile cluster of vesicles

internal motion, nor do they move from place to place. On the other hand, vesicularly disintegrating and swollen crystals of earth exhibit both internal motion and change of position. Crystals of earth which are not heated or swollen are electrically neutral, the vesicles of *earth* and *coal* crystals that detach themselves from the agglomeration of vesicles have a positive electrical charge and move vigorously. Large units of vesicularly disintegrating crystals of earth or coal are usually electrically neutral. What is the reason for this?

In the crystal, the particles are not swollen and the electrical energy is bound. Swelling or heating to incandescence breaks down the mechanical cohesion of the particles in the matter, releasing the energy. Otherwise the vesicles emerging from the crystal could not have an electrical charge. The relative movement in place of the vesicles can thus be explained as the result of electrical force fields of the individual vesicles acting on each other. If preparations of earth and coal which have been cooked or heated to incandescence are observed long enough and thoroughly enough, it is frequently seen that two vesicles dance around each other until a third vesicle comes near. Once it has approached to within a certain distance, one or the other of the two vesicles moves away from its partner and begins to dance around with the third, newly arrived vesicle, and so on.

If electrically charged vesicles of any type are held together by lecithin or gelatin, then, regularly, instead of the vesicles moving freely in space relative to each other, the agglomeration *crawls in a coordinated manner* and also expands and contracts as a system. It must therefore be assumed that the locomotion of the "vesicle heap" is brought about by the interaction of the electrically charged individual vesicles *inside* a mechanical boundary of, for example, lecithin. The vesicles are compressed into a space and they continue to move relative to each other within this space; but, as a result, the whole mass acquires intrinsic movement. In the agglomeration of vesicles there is an antithetical relationship or contradiction between the vesicles bouncing on and off each other inside the cluster and between these vesicles and the enclosing boundary; i.e., the surface tension. The electrical charge of the units within the boundary thus causes expansion and contraction. If, at a point on the periphery of a moving cluster of vesicles, the forces of repulsion acting on the individual vesicles predominate, then the pressure from within must increase. This "internal pressure" may even be *greater* than the resistance exerted by the boundary, in which case *stretching* occurs because the surface tension exerts less force than the internal pressure. But expansion leads to stretching of the surface, which in turn causes the surface to exert a new counterforce. The more the internal pressure increases the surface, the greater must be the tendency of the surface to contract again like a stretched spiral spring. It is not possible at present to say what effect this process has on the electrical charges inside the cluster. However, there is no doubting the antithetic effect of surface tension and electrical forces inside the cluster as the reason for the movement.

Pauli has demonstrated experimentally that the colloidal particles of protein enter into an intimate relationship with the solvent; namely, water. He explains this by stating that the water penetrates the protein and produces swelling. In Pauli's view, electrically uncharged particles of protein always swell less than electrically charged protein. The greater the ability of a protein particle to swell, the more stable it will be in solution. Protein particles that have swollen just a little or not at all also form very unstable colloidal systems; i.e., they flocculate easily. I would prefer to turn Pauli's explanation around and say: *the more a protein substance can swell, the more readily it can build up an electrical charge, and the more stable it is in colloidal solution.* The less readily a protein substance swells, the smaller will be the electrical charge, and the easier it is for the protein to precipitate out. There are certain advantages to be gained by turning Pauli's explanation around. For example, it indicates how one can solve the difficult problem of why the swelling of a particle causes the buildup of an electrical charge.

The relationship of the colloids to the gel is also very important. If dry gelatin is combined with water, then swelling takes place as water is absorbed, forming a gelatinous mass called a gel, as distinct from the sol of the usual colloidal solution. This biochemical finding is confirmed by our experiments, which have shown that if we combine protein substance with water or potassium chloride alone, we get a different result from that obtained when we add gelatin. In the first case, no amoeboid structures form; in the second case, the protein colloid has other properties which are closer to those in life. When lecithin and cholesterin are added, structure-creating forces are generated which, together with the earlier forces, create amoeboid structures. The gel, thus, probably embodies considerably more of the functions of life than the simple sol. The force by which the solvent is taken up by the gel is very great. During swelling, the volume of the substances increases markedly. Swelling occurs in all lyophilic colloids; e.g., protein. If the colloidal swelling takes place in a closed space, which prevents the increase in volume, very strong pressure is exerted. This is referred to as "swelling pressure." In order to squeeze the solvent out of the colloid, the same amount of pressure must be exerted on the colloid. The swelling pressure can often attain values of several atmospheres.

These facts explain some of the important phenomena in our experiments. There are other substances which can form colloids, such as metallic colloids, but the substances that make up organic life differ from the latter in that they also exhibit swelling pressure.

With the known biochemical facts in mind, let us now proceed one step further in our analysis of the problem of the mechano-electrical leap. For the particles to become electrically charged, it is first necessary for swelling to occur; i.e., for the particles to move apart. The second requirement would be for the expansion of the colloidal particle to be limited by the formation of a surrounding border; namely, a membrane. There would then be an antithesis between the internal pressure—that is to say, the pressure acting from the center of the swollen particle toward the periphery—and the opposing force exerted by

S = Surface tension
I = Internal pressure

the membrane toward the center. The swelling pressure of the colloid can only become an internal pressure if it is counteracted by the surface tension of the membrane.

The opposition between swelling pressure and surface tension is of eminent importance for understanding the division of cells into spherical objects. I will discuss the relevant facts elsewhere.

## INDIVIDUAL FUNCTIONS AND INTEGRAL FUNCTION

It is first of all necessary to deal with the objection that might be raised by researchers such as Hartmann. The amoeboid structures that are formed by heating certain substances exhibit motion—above all, expansion and contraction—as well as locomotion. One might object that these are not living organisms at all but merely surface tension phenomena of a purely physico-chemical nature. Let us see what Hartmann has to say:

> As is generally known, living systems are distinguished from inorganic systems by the fact that they possess highly complex organic compounds (above all, thick proteins). But if we attempt to distinguish between living bodies and non-living bodies on the basis of their chemistry, we find that chemical definitions get us nowhere because, *as things stand at present, there are no discernible chemical differences be-*

tween *living and non-living organisms*. And although at first sight it seems an easy matter to distinguish reliably between living or non-living organic bodies and inorganic bodies, if we consider the problem more carefully we see that such a distinction is fundamentally impossible. The chemist in his laboratory is able to synthesize chemical substances which are identical in all respects to the important organic substances such as carbohydrates and proteins that occur in living cells, yet these synthetic substances show no characteristics of life. Thus, given the present state of research, it is not possible to characterize and define living systems on a chemical basis. It is even less possible to find any fundamental physical differences between living systems and inorganic systems. Living systems are colloidal bodies and in this respect their properties are identical with those of non-living and inorganic colloids; research into the colloidal properties of matter in the last ten to twenty years has in fact provided some surprising explanations of many apparent peculiarities of living systems. An appropriate physical and chemical definition of life could really be attempted only if biology, instead of just being about to embark on nomothetic research, had already achieved its objectives, and if all the interrelated physical and chemical processes of the living organism were known.

Hartmann's statement forces us to take a serious look at the phenomena that we cannot help observing when non-living substances are heated and as a result take on certain characteristics of life. On the one hand, too sharp a line is drawn between the sphere of life and the inorganic sphere; when one thinks of life, one involuntarily thinks of something completely different from lifeless matter. Assuming that life is formed from non-living matter, we forget that there must be intermediate stages at which it is impossible to decide whether, for example, motion is due to "mechanical surface tension" or "spontaneous life." Yet there is no question that living movement and the pseudo-living movement of a drop of oil prepared for study are based on the same laws of surface tension and internal pressure of colloidal substances. The difference is that, in contrast to non-living objects, which may move as a result of surface tension, life, as it unfolds, possesses the property of *development*. It will never be possible on the basis of chemical or physical processes alone to distinguish "living" from "non-living," because physical and chemical functions are common to both. The only way to make a successful distinction between the two is not to deny the common aspects but to find those properties of life that *distinguish* it from the non-living sphere. The bion experiments have shown that the difference is not constituted by the *addition of something new* in living matter that makes it alive. Instead, the difference lies in a *special combination of functions* which are found singly in non-living matter as well. Let me recapitulate: mechanical surface tension and internal pressure occur in non-living as well as in living matter, just as the particles may also be electrically charged in both. However, *the specific combination of the rhythmic alternation of tension, charge, discharge, relaxation, renewed tension, charge, etc., is the fundamental distinguishing characteristic of life.* It is a new function and yet in its component parts it is not different from inorganic functions. The combining of inorganic functions into a *new type* of function might be sufficient explanation of the specific basic functions of life such as metabolism, division, reproduction, cohabitation, etc.

On the morphological differences, Hartmann has the following to say:

> Here we have such a fundamental, concrete difference between living and inorganic systems that one is inclined to assume that living systems can in principle be easily and reliably identified on the basis of their morphological peculiarities; i.e., their shape. A vertebrate, a fish, an insect, a phanerogam, or a fungus is a morphological object of such a characteristically distinct type that it naturally seems easy here to distinguish such systems from inorganic systems. But not even morphological characteristics can be relied on entirely, for there are animals, such as amoebae, which, morphologically, are virtually indistinguishable from any

liquid colloidal mixture, and there are plants, such as small bacteria, that are indistinguishable even from the smallest granules or droplets in dispersed systems (colloids). Thus, any attempt to characterize life on a general morphological basis seems to fail.

The only hope is to find the *unitary function* which can of course be broken down into individual physical and chemical functions which occur in inorganic matter as well but which do not function there as an integral whole. Until now, the point had been missed that the *unitary function does not in any way conflict with the totality of the individual functions.* This provided the vitalists—e.g., Driesch—with the basis for explaining the unexplainable in metaphysical terms. *The task of natural science can only be to determine the function which combines the individual physical, chemical, and mechanical functions into an integral function.* Therefore, we may say that living matter does not differ in any way from non-living matter as regards individual physicochemical functions. What is specific to life is that the individual functions are brought together into a unitary function governed by the two fundamental tendencies of *expansion (toward the world)* and *contraction (back into the self)*—as well as by the dialectical alternation of mechanical and electrical functions. Therefore, no matter how much I have agreed with Hartmann's statements up to this point, I find the following conclusion incorrect. He writes:

> If it were possible, with the present means of research, to understand and to analyze fully the chemical-energy processes taking place during so-called stimulus phenomena, it would not be necessary to use a separate term for this group of vital processes, which are now referred to as stimulus phenomena and which in many cases also play a role in the psychic sphere of higher forms of life; because then we would merely be dealing with more or less insignificant physiological fluctuations in the stationary processes, i.e., in the general metabolism.

It is clear from the foregoing that no analysis of the individual functions, no matter how detailed, can succeed in explaining the functioning of the whole that characterizes life. *The leap from the sum of the individual functions to the unitary functioning of all these individual functions must be grasped completely.* Thus, a colloidal mixture cannot be defined as possessing life if it merely exhibits the physical and chemical functions characteristic of life; instead, it can be defined as living only if these functions are combined into an organismic unity wherein all the elements of each individual function are incorporated into the unified function.

Metaphysical idealism, which treats organic systems as complete entities, has hitherto contrasted with mechanistic materialism, which analyzes the individual functions of the same systems; the two were absolute and irreconcilable opposites. The dialectical reconciliation of this contrast between totality and detail provides a satisfactory solution to the question of what is life by eliminating the metaphysical principle of the "beyond" as a means for explaining the whole.

In order to appreciate the miracle of life and the way it functions, we do not require any metaphysical explanation, however exalted. It seems to us miracle enough that we can establish an unbroken line from psychic excitation or gratification via the antithetical drives of pleasure and anxiety and the physiological antithesis of parasympathetic and sympathetic, via the chemical antithesis of potassium and calcium (lecithin and cholesterin), to the purely mechanical-physiological antithesis of internal pressure and surface tension as well as electrical charge and discharge. The continuity of this line from the simple inorganic vesicle to the highly complex system of psychic functions in human beings is in no way broken by the fact that it exhibits fundamentally different characteristics and complexities in the various developmental stages of function.

# 8

## An Error in the Discussion of "Spontaneous Generation"

A thorough and detailed discussion of the theory of spontaneous generation upon which so much attention is focused by these experiments is still not possible, because we need to know much more about the field in question. However, it is necessary even at this stage to point out an error that has crept unnoticed into earlier discussions on the question of whether life forms from life germs or through "spontaneous generation."

Around the middle of the seventeenth century, the Dutch scientist Leeuwenhoek discovered infusoria. By pouring water over substances such as hay, straw, pepper, and all kinds of spices, he found that the water eventually became filled with some strange animalcules. The question naturally arose as to how these very small organisms got into the infusions. The first information was provided by Richard Goldschmidt in a report in his book *Die Urtiere:* "The organisms were so small and seemed to have such an extraordinarily simple structure that it was very easy to believe that they had suddenly originated from some non-living matter. There had been no living organisms in the substances used to prepare the infusion—that is to say, in the pepper or in the hay—but suddenly life was present, and there seemed no other possible explanation than that the organisms, which had so suddenly appeared, had formed from these substances. Non-living matter had thus generated living matter, an interpretation which did not meet with the slightest objection."

At about the same time during research on intestinal worms and other parasites in animal bodies, it was found that certain

## An Error in the Discussion of "Spontaneous Generation" 141

flies developed from larvae. This seemed to deliver a severe blow to the theory of "spontaneous generation," replacing it with the "germ theory"; i.e., that all life develops out of life germs and not from non-living matter. Thus, the germ theory and the theory of spontaneous generation were at direct odds with one another. The experiment carried out by the English physicist Tyndall was regarded as an allegedly perfect refutation of the theory of spontaneous generation. Goldschmidt writes that the experiment was "particularly enlightening." According to Goldschmidt's report, Tyndall took a number of bottles, filled them with decoctions of fifty-four different substances such as meat, etc., heated them to over 100°C, at which temperature no living organism can exist for long. He then fused the bottles shut and took them with him to Switzerland. During the journey six of the bottles broke, and when he examined their contents he found something very strange. Despite the heat treatment, all the decoctions contained living organisms. He divided the remainder of the bottles into two equal groups. He carried one group to a glacier, where he broke the necks off the bottles and let them stand for several weeks in the pure air. He took the other bottles to the hayloft of a house, where he opened them. A short while later, he examined *all* the bottles. The bottles which had stood in the hayloft, like those that had broken en route, were full of all possible forms of life, while not a trace of life could be found in the bottles that had been exposed in the pure microbe-free air of the glacier. "This test shows," Goldschmidt wrote, "that animals can only develop in such decoctions in impure air."

An obviously emotional resistance to the idea that living matter could develop from something inanimate prevented Goldschmidt from seeing how incorrect this argument was. What did the physicist Tyndall do? He prepared decoctions of substances and after a certain time found *living organisms in hermetically sealed bottles,* just as I did in my experiments. He took some of the bottles, exposed them on a glacier, and found no living organisms in them. From this, he incorrectly concluded that the living organisms he had found in the sealed bottles had been introduced through impure air from outside. *This conclusion was unjustified.* The only correct conclusion to be derived from the effect of the glacier air would perhaps have been that these organisms *cannot survive* in glacier air. The glacier-air experiment does not in any way provide a positive explanation of how living organisms can get into decoctions in bottles that have been sealed. It is much more probable that the living organisms were formed by the decoctions and then *were killed off in the glacier air.* One should not be so quick in denying the possibility that living organisms are formed from non-living substances. There is a strict rule that positive scientific results should always be checked and verified. Just as much care should be taken when denying facts. There can be no doubt that living organisms are found in decoctions. Right from the start, it was wrong to ask "How did the organisms get into the sealed bottles?" since it is known that no living organisms can penetrate glass containers that have been sealed.

The pioneering experiments performed by Pasteur seemed to provide further support to those who rejected the theory of spontaneous generation. Pasteur proved that air which is completely "germ-free" never produces living things when brought into contact with nutrient substances. Goldschmidt remarked in this connection: "For the scientific world this *completely answered* the question whether spontaneous generation exists, and it further showed that if living organisms suddenly occur where previously there were none, they developed from the germs that are contained everywhere in dust or air." But this is far from clear proof; *so-called germ-free air does not contain the very fine particles of dust which, upon swelling, yield living organisms.* Pasteur collected dust from the air and found that there are a number of microbes that are suitable under favorable conditions—namely, *in a moist environment*—to re-create *small organisms.* This statement is already prejudiced because the *nature and the origin of these microbes* still remained unexplained. *The microbes are either organisms which are, so to speak, dormant and which develop into another form of organism—namely, bacteria or protozoa* (in which case one would like to know what exactly is the nature of these immotile organisms that we call "germs," and as far as I know, no such knowledge is available at present)—or *the germs are very fine non-living substances which convert into living organisms following moisture-induced swelling* (in which case this process of transformation would have to be proven by scientific experiment). The fact that organisms develop out of dust-laden air whereas none develop from dust-free air in *no* way indicates that life cannot form from inanimate matter. The fine particles of dust could in fact be the very germs of the lifeless substances from which living organisms develop, depending on the definition of the term "germ."

In a dried state, the dust may still contain bacteria, which come to life again when brought into contact with moisture. Are these dried-out bacteria still living organisms or merely "dust particles"? The germ theory thus need not be at variance with the theory of spontaneous generation if one merely accepts the premise that non-living matter can change into living matter and vice versa. The supporters of the germ theory, as it has been defined so far, must still answer these questions: What characterizes the germs as latent organisms? What distinguishes them from non-living matter? How do they develop into motile life forms? *Pasteur's experiment thus does not disprove the theory of spontaneous generation but merely discloses the effect of dust particles in the air.* If one observes decoctions after ten or fourteen days and finds living organisms, it is still conceivable that the organisms had entered externally. However, a decoction taken immediately from the sterilizer, put under the microscope, and found to contain living organisms would seem to exclude this theory. Otherwise, it would be completely impossible to sterilize anything; even germs require a certain amount of time to develop. In addition, germs should be completely killed off by a long period of heating at temperatures in excess of 100°C. These contradictions in the germ theory need not exist if one stopped regarding the germs as "complete organisms."

I should point out an inaccuracy in the concept of spontaneous generation. Spontaneous generation can be taken to mean that, at some time in the past, life formed from non-living matter and was capable of reproducing itself. Spontaneous generation, however, can also mean that now, *continuously, in every minute and every second, life is developing from non-living substances.* If this can be proved beyond all doubt, it would naturally not contradict the development of life from germs, because spontaneously generated life could maintain itself through reproduction.

However broad the perspectives of such a view may be, we must, for the time being, restrict ourselves to verifying precisely the process by which inorganic matter is converted into organic matter. Neither the infusions prepared by Leeuwenhoek nor Tyndall's experiments nor those performed by Pasteur mention this process of *transformation,* because these researchers did not observe their preparations *continuously.* There are a large number of *developmental stages* between the inanimate particle of soil or the individual plant fiber and the observations of completely organized microbes or protozoa. There are *preliminary stages of completed organisms* and there are stages in which it is difficult to decide whether the clump of earth or the plant fiber is non-living or alive?

Some of the experimentally generated types of organisms can be described as complete life forms. This expression cannot be used so readily to describe other types which should be referred to as preliminary stages or developmental stages of living matter, as in the case of the plasmoids becoming crystals.

The swelling of a decomposed plant fiber or of a crystal of earth is definitely still a mechanical, non-living process. When a taut vesicle detaches itself from a plant fiber, this is presumably already a living process. But, for example, the formation of a utricle from a swelling particle of earth could be regarded either as a mechanical or as a living process, depending entirely on the definition of the concept "living." Therefore, it seems important to me when verifying and reproducing the stated facts to trace in particular the *details* of how inorganic matter is transformed; i.e., the process of swelling, vesicular disintegration, the formation of vesicles, etc. It must be assumed that there are no boundaries between the plant world and the animal world, between inorganic non-living matter and living matter. This requires that we abandon the static, mechanistic way of thinking and attempt to comprehend the process and dynamics of the functions.

## SUMMARY

1. *The theory of spontaneous generation has never been definitively disproved.* Leading researchers regarded it as an inevitable consequence of scientific thinking.
2. The aim of the opponents of the theory of spontaneous generation has always been to prove that the formations in question can be killed at high temperatures. *The proponents of the theory of spontaneous generation have never been able to provide exact proof that spontaneous generation actually occurs.*
3. *The decoction was not examined immediately after it had been prepared.*
4. No attempts were made to culture the structures produced by the decoction.
5. The transitional stages from non-life to life were never discussed in conjunction with experiments.
6. It was not possible to carry out observations at 2000–4000×.
7. No exact research was carried out at temperatures above 180°C because people were so certain that any existing life was killed off that they completely overlooked the possibility of new life being generated at temperatures above the lethal limit.

# 9

# *The Dialectical-Materialistic Method of Thinking and Investigation*

## THE BASIC METHODOLOGICAL APPROACH TO OUR EXPERIMENTAL WORK

It would not have been possible to carry out the bion experiments if, in addition to employing the fundamental dialectical-materialistic approach, the work had not also been inspired by a certain attitude toward scientific research. Let me try to describe the main features of this attitude. Any scientist who ponders and scrutinizes the results must also have some appreciation of the atmosphere in which the work was conducted.

There is a world of difference between scientific work, which is concerned with arranging, standardizing, and detailing already known facts, and thus keeps to known regions, and research, which for the time being must dispense with such comforting security. A fundamental characteristic of the latter is that the researcher is uncertain and doubtful about what he thinks he sees. Any scientific discovery represents an advance into unknown territory and usually comes more or less directly into conflict with well-known *theories*, with subjective interpretations of verified facts. Thus, in addition to coping with the unknown, the second type of research also has to grapple with established views, to refute them, to confirm them on a different basis, or to make different use of them.

It is clear that scientific work conducted, for example, in the public hygiene institute, which is organized and pursues a set

## Dialectical-Materialistic Method of Thinking and Investigation 147

course, will use other means to achieve its goals than work that is being conducted in uncharted territory. Such routine work usually automatically eliminates anything that departs from the normal course of procedure. It avoids using any techniques other than those that have proved to be essential and reliable for performing certain tasks. It is perhaps unfortunate, yet inevitable, that new facts uncovered by any scientific advance simultaneously block the way for further advances because the newly discovered facts have yet to be organized. For example, Pasteur's discovery of sterilization was certainly *the* major scientific advance of the last century and it placed all medicine on a new footing. But as du Teil has already stressed in his lectures, this discovery barred the way for processes to be investigated *above* the known maximum sterilization limit. It is therefore unavoidable that pioneering research will at first be unpopular and encounter obstacles. It throws established scientific knowledge into confusion. One can safely say that it takes a certain amount of recklessness and foolhardiness to conduct research in new areas. Let me give some concrete examples:

Since bacteriological technique has involved the killing of life, it is only natural that research into microbes should make use of the practice of killing and staining the killed material. Staining the preparations reveals structures clearly and preserves them for continuous study. In direct contrast to this, however, it was one of the principles of our work that we study only living, motile matter, because it is the *changeability*, the *functioning*, and not the static element, not the structure, which are of primary interest.

Provided one has good critical self-control, one can risk making "unscientific" leaps which would be utterly forbidden in usual scientific work. For example, it was essential at the beginning of my work, even if it was not intended, to mix the substances together *without sterilizing them*. If I had started off with a high degree of sterilization, the distinction between the relatively immotile nature of non-sterile matter and the motile nature of extremely sterile matter immediately following production of the preparation would not have revealed itself. In the first experiments, the swelling and, in particular, the culturing of extremely sterile crystals of coal seemed even to me to be a little "mad." Yet it was precisely this leap into a highly unscientific procedure that made it possible to explain the spore theory. Such leaps are only valid if control experiments are carried out afterwards. However, scientific research is often stifled by an excessive amount of such controls.

As an example of this problem, during the bio-electric experiments my first co-workers spent weeks and months establishing the phenomena that result when the wires leading to the oscillograph are mechanically agitated. The resulting fluctuations were in the order of one to a maximum of five millivolts. The experimenters were so engrossed in carrying out this control that they, for example, totally overlooked the 20, 30, and 50 mV fluctuations caused by tickling the palm of the hand. I was very surprised when one of them, upon examining the first trace of an excitation fluctuation, grew enthusiastic about the visible cardiac spikes. These were about 1mm high, whereas the spikes of the excitation fluctuations were about 2 cm high. This phenomenon troubled me greatly. I tried to explain it, I thought correctly, by assuming that even the most conscientious scientist is so tempted to give in to playfulness and foolhardiness that he must protect himself from this weakness by relying too much on equipment and by performing excessive controls. Both these ploys can have harmful effects. Some scientists refuse to consider a result unless they have an extremely accurate description of the apparatus used. To my mind, this is an exaggerated attitude. It is of course extremely important, when passing an electrical current through a preparation, not to diagnose immediately the movement observed under the microscope as cataphoresis. First of all, it is necessary to possess an apparatus with which one is fully familiar and which permits the *current to be reversed.* In this way, cataphoresis can be distinguished from a streaming of the fluid. But the apparatus should not be more important than the phenomenon which one wishes to understand. In the course of the

## Dialectical-Materialistic Method of Thinking and Investigation 149

present work, I had to learn to keep an open mind on some very questionable—indeed, very improbable—phenomena, and at first without carrying out any control tests, so as to let them create their own impression on me. It never fails to surprise me what results one can arrive at by consciously adopting this highly unscientific approach, always providing that the playful leaps are later verified by strict controls.

I should mention another extremely regrettable attitude. A researcher discovers a phenomenon—e.g., movement in India ink—describes it, makes it known. The phenomenon is named after him and is called Brownian movement. Brown believed he had discovered a living phenomenon. Physicists, however, explain the phenomenon as a purely *physical* manifestation brought about by molecular motion and, contrary to the view of its discoverer, Brownian movement is now established as a concept with which one "explains" *any* microscopic movement that is not clearly and obviously associated with living matter. It is not pleasant to hear an expression used over and over again when one is demonstrating the existence of other totally unrelated phenomena. One physicist admitted to me that he was teaching Brownian movement to junior high school students although he had never seen it himself. Romeis's handbook on bacteriology does not even mention Brownian movement, as it belongs, after all, in the realm of non-life and will forever remain there. Because of the existence of this attitude, I made it a methodological principle of my work *to adopt any technical achievement of scientific research and to record accurately every experimental observation, but, so as to avoid confusion, to ignore all theoretical interpretation for the time being.* To eliminate problems, I explicitly asked my co-workers to apply only their technical skills and their knowledge of the facts, but, in the course of their work and in their dealings with me, to forget any theories or interpretations that they had learned. This made it possible to carry out the experiment with incandescent soot, allowing it to swell in broth + KCl, and even to culture it. In fact, it turned out that this so-called physical motion [Brownian movement], which should *always* be present, is absent when the conditions are modified. The particles of coal and soot lie quite still and only begin to move after a period of swelling. The movement continues for several weeks or months before it finally ceases.

When someone breaks through into what seems an improbable area, he becomes keenly aware of the phenomenon of lazy thinking in scientific quarters. Thus, in the second type of scientific research, one has to do more than just grapple with facts and problems and fight against or refute traditional and often false views. It is absolutely essential to believe that one's own views are accurate and correct, yet, at the same time, one has to overcome the inevitable and torturing doubts about one's own work. While these doubts can be very fruitful, they are quite often the reason that work is abandoned at an early stage. I heard one university teacher say that he had once worked along the same lines but had given up because of excessive difficulties that were put in his way.

So much for the fundamental attitudes that are essential for such work. The main rule is: do not automatically believe in anything; convince yourself of something by observing it with your own eyes and, having perceived a fact, do not lose sight of it again until it has been fully explained.

The methodology of dialectical-materialistic thinking and working, which we consciously apply, is equally important. A fundamentally dialectical-materialistic approach requires that the organism be examined as it is; that is to say, that life be studied in the living state. This approach is diametrically opposed to the mechanical one in which, for the sake of reliability, living objects are killed in order to study life in the dead organism, a procedure that is bound to result in a mechanical view of life. Furthermore, the individual living object, or even a detail of this object, should not be studied as an isolated phenomenon. The basic dynamic principle of life governs all life; i.e., the organism as a whole and every individual part of the organism. If scientific research is to be truly productive, it should be continuously motivated and

## Dialectical-Materialistic Method of Thinking and Investigation 151

guided by the need to view the whole without losing sight of the detail. Mechanical concepts of life must of necessity be methodologically defective; they rely on the synthetic movements of the living substances becoming more complex and perhaps giving rise to life. The important thing about life is not, however, the complex substance but the complex function. Concepts such as biogen, molecule, energid, etc., are only practical aids to understanding. They bear no relationship to any facts. They try to substitute the action of "substance" for the understanding of function. They tell us something only about the mechanical-chemical process, but they become metaphysical when called upon to explain function. Since these concepts are nothing more than aids to understanding and not actual facts and since they do not encompass any functions, they are often more of a hindrance than a help to practical research. There is only a new X, which becomes God as long as the function itself is not tangible and reproducible. What is unfortunate about the chemical-mechanical concept of life is that one tries to arrive at the whole from the part by *adding together details,* instead of *seeking the function of the whole in each individual part.* From our basic methodological standpoint, there is no difference between the plasmatic streaming of an amoeba, which one can see, and the vegetative current which one experiences in certain states of excitation. The function of a tree cannot be explained by defining the chemical composition of the cellulose. The leaf branches out in exactly the same way as the limbs of a tree, the individual branches of the supporting structure, and the veins of the leaf. A unity dominates the whole.

Physiologists, for example, describe the pathways and directions of impulses, believing that this explains the impulse itself. Explaining the observed phenomena in mechanical-material terms and dividing the entire organism into individual parts prevent us from grasping the biological function of the whole. Physiologists also separate emotions and moods from the vegetative functions; e.g., "joy brings a blush to the cheeks" (an idealistic-metaphysical view of the event), or "fear is accompanied by sweating," or "the function of probably every organ innervated by the autonomic nervous system would be influenced by some type of mood."

The function of the organism is viewed in sociological terms. The brain is the "central agency," the "controller of the organism," rather like the ruler of a country. But the brain is a phylogenetically recent structure; there are organisms that have no brain at all. The *living organism* is more primitive and older than the brain and therefore constitutes origin and center in the functional sense. Life is possible without a brain, but a brain cannot exist without vegetative life. Such conceptual errors are caused by the intellectualistic projection of one's own perception of functions onto reality. For decades, the view prevailed that the vegetatively innervated organs do not transmit sensation. On the other hand, it is already possible to measure the state of the ego as a reflection of the vegetative excitations active at any one time. A person feels the contraction of the sympathetic nervous system directly when he stands close to the edge of a sheer precipice. We experience pleasure as expansion, as stretching, widening. Since organ sensation and vegetative life are identical, it necessarily follows that a worm must be aware of itself, because vegetative current is identical with feelings of pleasure and unpleasure.

These comments were necessary because they characterize to some extent my fundamental approach. Admittedly, measurements and replicate experiments still have the last word in science. But when I see an amoeba stretching and the protoplasm flowing in it, I react to this observation with my entire organism. The identity of my vegetative physical sensation with the objectively visible plasma flow of the amoeba is directly evident to me. I feel it as something that cannot be denied. It would be wrong to derive scientific theory from this alone, but it is essential for productive research that confidence and strength for strict experimental work be derived from such involuntary, vegetative acts of perception.

## THE DIALECTICAL-MATERIALISTIC LAW OF DEVELOPMENT

*The Conflict between Mechanistic and Vitalistic Views in Biology*

Compared with the mechanistic materialism of the Büchner type, as well as with the no longer absolute but dialectical idealism of Hegel, the materialistic dialectic represents an infinitely fruitful, and little exploited, means of enriching scientific thought. Of course, it is not enough merely to vaunt the materialistic dialectic as the better method or to discuss it in abstract philosophical debates; i.e., to pursue, as it were, a "method in itself." Two points are absolutely essential here: the dialectical-materialistic method should be tested in a concrete situation on actual scientific and social problems; and it should be used in a new way to solve these problems, to *prove* its superiority, to show that one *does in fact* discover and achieve more with this method than with the mechanistic methods.

Let us briefly summarize the principles of the dialectical-materialistic method:

Mechanistic materialism asserts that development takes place in a "causal sequence"; that is to say, that all phenomena have a cause, that this cause is itself the result of an earlier cause, and so on. However, this method does not really explain how event B derives from event A. Therefore, one has to assume a "development principle," which itself needs to be explained. After all, why is there such a thing as development? The mechanistic method can only answer this question by having recourse to a "force" inherent in matter which is nothing more than the "spirit" referred to by metaphysicians. But where does the influence-exerting force or spirit come from? Dialectical materialism, on the other hand, states that development comes about from the presence of opposites within matter which cause an antagonistic contradiction. This contradiction cannot be solved within a certain situation. Therefore, the opposites force a *change* in the situation, and something *new* is formed. This *new something*—

formed through the resolution of the contradiction—develops new contradictions, which in turn force further solutions, and so on. Thus, everything is in a constant state of flux. Nothing is separate and absolute, everything interacts.

In the mechanistic view, the opposites are absolute and irreconcilable. In dialectical materialism, opposites are viewed as identical and as a consequence one can develop out of the other. Hate is not merely an opposite of love, it can develop out of love; much conscious love is unconscious hate, and vice versa.

From the age of enlightenment onward, mechanistic science conveyed the concept of development; things are not only eternal but are also in the process of developing. But mechanistic science linked with a *Weltanschauung* represents the standpoint that the development is a gradual process and *nothing else*. There are no *sudden* changes. The materialistic dialectic, on the other hand, recognizes that gradual development can become sudden development, that evolution prepares the way for sudden change in development. Scientific research very often disproves mechanistic philosophy; e.g., when it correctly recognizes the sudden changes in the course of development.

Mechanistic as well as idealistic philosophy denies that it is possible for quantity to develop out of quality, and vice versa. In contrast, dialectical materialism asserts that not only can quantity convert into quality, and vice versa, but that this change-over is one of the fundamental principles of any natural process.

Scientific knowledge advances much more slowly than the urge for knowledge. Natural philosophical theories arose from efforts to proceed ahead of acquired knowledge, grouping it into an overall philosophical view of natural events. Two main directions of thought stand out: *mechanistic* materialism and *vitalism*. A researcher does not arrive at certain general theories merely on the basis of his own findings and those of others. He usually approaches matters consciously or unconsciously with a general philosophical view. The theories that he develops will also depend on his *Weltanschauung*. One cannot try to analyze the boundary

## Dialectical-Materialistic Method of Thinking and Investigation 155

between the psychic and the physical without having some knowledge of the basic scientific theories. I shall therefore briefly summarize the difference between the mechanistic-materialist *Weltanschauung* and the vitalist *Weltanschauung*.

Mechanistic theoreticians seek to understand the organism as a machine. They investigate the energy processes, the chemistry, and the physiological laws. They assert that the same laws apply to organic as to inorganic matter, the only difference being that in the former the laws are fundamentally more complex and therefore more difficult to comprehend. But, in principle, the mechanists believe that it is sufficient to know the physical, chemical, and physiological processes in order to be able to understand life.

A vitalist starts out from a much more advantageous critical position because of the poor state of knowledge available. He maintains that, aside from the causal relationships between the parts in organic matter, it is also necessary to understand the purpose and function of the *whole*. Development is not arbitrary but takes certain forms which cannot be explained causally. The development of the individual parts into the whole, into species, etc., is determined by a goal, *telos*. This goal is survival-oriented, to preserve and shape the whole and to propagate the species.

The mechanist maintains that no essential distinction can be made between the inorganic and the organic as regards the relationship of the parts to the whole. In crystals and chemical compounds, for example, no part can be removed without destroying the whole. The only possible difference is that in the organic the relationship of the parts to the whole is simpler to analyze.

The vitalist counters that the *sum of the parts does not yield a functioning organism*. The whole is dominated by a purpose, an idea. As we know, this is the same objection that the humanities constantly level at science, that axiologists level at materialists.

The mechanist is able to refer to Köhler's "physical forms"

(*physische Gestalten*), which prove the wholeness of inorganic systems.

However, the vitalist knows how to defend his position. He says that the existence of hydrogen and oxygen and also their combination into water can be proved chemically. But $H_2O$ possesses completely new properties that were not present in the H or in the O. These properties are "irrational" and can never be understood in mechanistic terms. Qualities ("green," "red," "liquid," etc.) are, specifically, newly formed properties and therefore not accessible to understanding. (Form and kind pose similar problems.) The physics of conscious processes may show vibrations, but it does not explain the color "green."

A mechanist like Max Hartmann would reply that science is not interested in explaining the irrational, that the irrational portion of existence lies outside the sphere of science. Purpose explains nothing because it itself needs to be explained. The purpose principle is a guide only to basic research, permitting it to penetrate into otherwise uncharted areas.

A vitalist like Driesch would say that the purpose principle is necessary. Though each cell is a part, it can develop into the whole, indicating the "prospective potency" of organic matter. In Driesch's words, an organism is a "harmonious equipotential system." This is certainly not true of a machine. No machine can be produced out of a single screw.

A mechanist would counter: Each cell contains all the chemistry to permit differentiation. It is necessary to know the fine colloidal structure. While there is certainly a gap in our knowledge here, this does not prove that it is impossible to find an explanation.

The vitalist refers to heredity and mitosis in explaining how each egg can produce the whole organism. From the mechanistic standpoint, one would have to regard the egg as a three-dimensional machine. However, division from *one* primary sex cell excludes the possibility of a machine, because a machine could not divide so many times and still remain whole.

The mechanist points to chromosomal division, to the equal

## Dialectical-Materialistic Method of Thinking and Investigation

division of potencies. The vitalist counters that cell division cannot be explained in mechanistic terms.

Enough of that. We could continue the squabble ad infinitum and still not resolve it. I have mentioned the main points merely to make clear our fundamental methodological approach, which is diametrically opposed to that of the mechanistic-materialist as well as the vitalist.

### The Three Dialectical Systems

Dialectical materialism denies the existence of a purpose principle in nature; i.e., the existence of a supernatural goal beyond the energy process that, according to the general view, determines development. But it also denies the existence of a mechanistic causal principle. If, in fact, we assume *one* material force which controls the development of any process, then at the same time we unconsciously reintroduce a metaphysical principle that acts beyond matter and energy. Science is nowadays dominated by the two above-mentioned principles of thought and interpretation; namely, metaphysical fatalism and mechanistic causalism. But, in rejecting these, dialectical materialism is faced with the difficult task of proving that it can solve contradictions and provide explanations which are not possible with other methods of reasoning. It must prove itself in a *practical* situation by solving, *through experiment*, in its own special way, hitherto insoluble problems. It goes without saying that the working method adopted by dialectical materialism is based primarily on its own fundamental principle that theory should be proven by practice and that theory and practice should be one.

The developmental mechanist and the teleologist will both say: "Fine. So all development proceeds in a series of opposites and contradictions. We understand that if force A enters into conflict with force B, something new, a third force C, results in which A and B are nevertheless still present. C comes into contact with D and the process continues in the same way. But what happens in the case of *organic development* when a germinal development takes place without the germ coming into contact with an external force? It is clear that, in this case, development proceeds from a principle contained *inside the germ*. But *what gives rise to this inner contradiction* which drives development from the center toward the periphery? Even this inner contradiction must itself have formed at some time!"

This critical question which the developmental mechanist and the teleologist ask is entirely justified. It is not enough for dialectical materialism to assert that a new third force arises from two conflicting forces. It is absolutely essential to grasp very clearly how the *inner contradiction comes about* and to understand its function. A review of the scientific arsenal of dialectical materialism forces us to admit that, so far, this question has neither been posed nor answered. Up to now, it has been difficult to find a place for biological development in dialectical-materialistic thought.

It is necessary to distinguish between three types of antithetical relationships in nature:

1. The *antithesis of systems (system antithesis)*. If a body system A encounters a second body system B—e.g., two spheres coming together—both must somehow change their direction. A *third* direction is generated. A ratio of the centripetal and centrifugal forces of heavenly bodies and the resultant, the orbits of the planets, is a typical result of a dialectical system antithesis.

2. The *dissociative antithesis*. If I move the positive end of a magnet toward a neutral piece of iron, the iron and the magnet form a *system* antithesis. This has a certain effect on the iron; its molecules undergo *inner* reorganization. On the side toward the magnet, the indifferent iron becomes negative, and on the side away from the magnet, it becomes positive. The effect of the magnet sets up two opposite directions in the uniform piece of iron. If we remove the magnet, then the antithesis is eliminated and the iron returns to its neutral state. The "environment" (the magnet) makes the iron itself magnetic; i.e., part of the environment is "inside the iron." In other words, we might say that an *external* contradiction or antithesis has become

## Dialectical-Materialistic Method of Thinking and Investigation 159

an *internal* contradiction. The pattern of the *dissociative* antithesis is as follows:

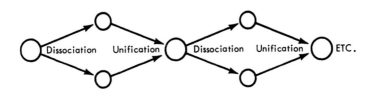

Magnetism, electricity, salt–base–acid, etc.

Here is another example: Positive and negative electricity, as well as chemical dissociation—e.g., NaOH + HCl = NaCl + $H_2O$—are also subject to the law of the dissociative antithesis. *In the process of dissociation, the tension of antithesis arises out of the neutral state, and this tension drives toward elimination of the antithesis; i.e., toward relaxation.* Tension and relaxation are thus not just successive independent processes, *but the process of tension leads inevitably to relaxation and, similarly, relaxation already contains everything that is necessary to lead to renewed tension.*

The salt $Na_2SO_4$ + $2H_2O$ gives 2Na + 2 OH + 2H + $SO_4$ during electrolysis. If I now combine the antithetical chemical compounds caustic soda and sulphuric acid, the unified neutral product $Na_2SO_4$ (sodium sulphate) and $2H_2O$ (water) is produced. Let us illustrate this interaction of tension and relaxation by considering the example of a steel spring. If I stretch a steel spring, the "impulse" in the spring to return to the relaxed state of rest will increase in direct proportion to the amount of stretching. When I release the spring, it returns by itself to the relaxed state. The difference between organic life and the mechanically tensioned lifeless spring is that *life can generate new tension by itself.*

3. The *genetic antithesis*. So far, we have only considered the alternation between unification and dissociation. A living cell dissociates before it divides, at first internally, through the process of mitosis. The copulation of two such internally dissociated cells signifies reunification of the antithesis. But the process of cell division differs from simple dissociation in that *it proceeds in a geometric progression*. A cell divides into two, these divide into four, the four into eight, sixteen, etc., in the following pattern:

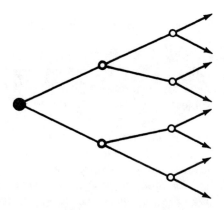

We shall return later to the problem of how genetically differentiated material later becomes reunified. Looking ahead, we can say that *the differentiation in the growth of a biological organism goes together with a constant combining of the differentiated biological units into a uniform function of the whole organism.* If we now consider the three cited fundamental forms of the dialectical function in nature, we find that:

The *system antithesis* applies in all areas of existence. There is no area in which a body is not opposed by some other body in some function or another, whether it is in the interaction of molecules or in the relationships between human beings or the heavenly bodies.

The *dissociative antithesis*, in the form of mechanical, chemical, and electrical processes, dominates inorganic as well as organic existence.

*The genetic antithesis is exclusively associated with organic existence.* The antithetical relationship between two bodies in space may be the cause of a dissociative antithesis (magnetic iron), in the same way that the dissociative antithesis can give rise to a genetic antithesis (cell division). All three types of antithesis occur in the world of living organisms. *The organic and the inorganic have in common the system antithesis and the dissociative antithesis, but the organic differs from the inorganic through the function of the genetic antithesis.* Copulation and procreation correspond to the unification of two antitheses into a third new entity. Propagation of the individual is at the same time dissociation of the single entity. The biological symmetry of organisms is a *direct* expression of the dissociative function.

Let us consider the fundamental processes of chemical dissociation and electromagnetism: Each dissociation or division of the single entity into opposites signifies a release of energy and at the same time the creation of a tension that seeks release. Any unification of antitheses represents a binding of free energy; in other words, a relaxation. Sodium chloride is neutral. When it is dissociated into sodium cations and chlorine anions, the result is antithetical positively or negatively charged particles that strive toward equalization; i.e., toward unification and neutralization. In the same way, the piece of iron is neutrally unified "without tension." If the iron dissociates under the influence of a magnet into positively and negatively charged particles, energy is released. An antithetical tension is formed. We recognize this by the fact that the previously indifferent piece of iron is now able to *perform work*. For example, it can now move another piece of iron. Similarly, the dissociated electrolyte is able to perform work by means of the ionic current, but the neutral salt cannot.

Let us now transfer this knowledge to the fundamental processes of biological functions. To begin with, we see that the sperm cell and the egg are energetically highly tensioned, antithetical systems which strive toward unification, toward equalization of the tensions, in the fusion of egg and sperm cell. The previously external antithetical relationship of the two is now replaced, following union, by an inner antithesis. The *division* of the fertilized egg begins as the expression of an energy-driven process of multiplication. The system antithesis of egg and sperm cell leads directly to the function of the genetic dissociation of development in a geometric progression. From the egg, first two, then four, eight, sixteen cells form.

It is obvious that all life must possess a *common basic rhythm*, a *common basic law* which it should be possible to detect in every detail of life. The division of the life process into biological, psychic, ideological, cultural, etc., areas is—and we should never forget this—an artificial classification made to satisfy our practical needs. How does this inner antithesis of a biological system which leads to genetic differentiation and to biological development, such as the fertilized egg, come about? Let us try to answer this question by, first of all, taking an example from an area with which I have become familiar through character-analytic work. We will consider how an external antithesis is internalized and how this internalization manifests itself in the course of development.

Disregarding complicating factors, a child that plays harmlessly with its own feces, for example, in the first or second year of life, is—with regard to the smearing of the feces—structured in a uniformly drive-affirming way. The drive to play with the feces then comes into conflict with the outside world's insistence on toilet training. We have here a basic example of the *formation of a system antithesis*. This antithesis now leads to a conflict between the child and its educator. Under the constant influence of restrictive education, the external antithesis changes into an internal one. The first result of this transformation is that the child becomes afraid of the educator and this fear conflicts with the impulse to play with its excrement. This fear is the starting point for the formation of an inner psychic conflict. The structure of the child becomes progressively divided into two parts: the

## Dialectical-Materialistic Method of Thinking and Investigation 163

impulse to smear the feces remains, but fear prevents the child from doing so. Something in the child, which we call the ego, starts to defend itself against the impulse and develops a moral imperative from its fear of punishment. Where, before, the child said to itself, "I am afraid to play with it," now it says, "I won't play with it because it's dirty." The external antithesis has become an internal antithesis. The unified impulse has become dissociated, split up. This internalized conflict becomes a source of "development" in terms of a genetic antithesis. The antithetical relationship of "desire-to-smear-the-feces" and "fighting-the-urge-to-do-so" leads to the point where the child now starts to prefer to draw or to paint. Out of the inner conflict, something new has formed which is made up of the two antithetical elements. The new something represents a unification of the antithesis. If the child now paints or draws, it unifies all existing demands, those of impulse as well as of inner moral imperative, because, in contrast to smearing feces, drawing is "clean." Painting and drawing are culturally and socially important and recognized activities. Thus, we see how in the product of sublimation, "painting," both the old system antithesis and the original internal conflict between impulse and morality have become unified into something new. The following diagram illustrates the process. Let us

leave aside the interesting question of how such a unification again becomes differentiated. What should be noted is that the process of dissociation and opposition, as depicted in the diagram, is not a static but a *dynamic* process *governing all forms of existence* that undergo development.

All development contains within itself the following functions united into *one* common function:

1. System antithesis.
2. Dissociative antithetical relationship or division (through internalization of the system antithesis).
3. Opposition of the divided forces.
4. Reunification of the divided forces, with subsequent progressive splitting up in a geometric series (only in the organic sphere).

Let us examine the foregoing statements in some fundamental areas. Organic life originates from inorganic matter. It contains within itself the specific mechanical, chemical, electrical functions of the inorganic. However, in the process of evolving, it develops its own laws, which remove it from the inorganic. The plant, for example, contrasts with the fluids in the soil in which it grows; the org-animalcule contrasts with the bion formed from the plant by sucking up, "eating" this bion; or man, who is a part of nature, contrasts with nature itself.

After it has become differentiated from the inorganic sphere (mainly through the action of eating), organic, vegetative life undergoes further differentiation with the development of consciousness. But vegetative life and conscious life, which derived from a *single* root, also contrast with each other (as, previously, vegetative life contrasted with the inorganic sphere) in the development of self-perception. Life perceives itself through the consciousness function examining its own origin. In the area of sociology, the process of dissociation and opposition is evident, for example, in religion and sexuality. They were originally identical because natural religion was nothing more than collective orgastic experience. With the invasion of private enterprise and [compulsory] sexual morality, the orgastic experience in society became differentiated. The sex-affirmative religion turned into a sex-negative, supernatural doctrine which now raged against its own origin. Religion has become the negation of sexuality, its antithesis, despite their common origin. In the new ideology which affirms natural sexuality, a further step is taken in this development: sexuality confronts and combines with religion in recognition of vegetative life.

In scientific development as well, we see that a new theory can emerge from an old one as a result of differentiation, following inner contradictions and subsequent opposition. Because the old science of psychiatry did not recognize the existence of the psyche, psychoanalysis, embracing it, split off and contrasted with the old psychiatry. Analytic psychology, on the other hand, contained in itself a contradiction between the clinical theory of repression and the theory of culture. From this arose (at first as an attempted solution within the psychoanalytic system of thought) the theory of the *unity of drive and culture* in the form of the *scientific orgasm theory*. An inner contradiction between the orgasm theory and the psychoanalytic culture theory led to a differentiation of homogeneous psychoanalysis into the sex-economic theory of the orgasm on the one hand and the death instinct theory on the other. In the further course of development, sex-economy formed an *antithesis* to psychoanalysis. At the same time, it represented a new concept that was able to solve many earlier contradictions between, for example, sexuality and work, drive and morality, nature and culture, etc.

After this digression, let us come back to a subject closer to the topic under discussion. One of the most striking examples is the antithesis of sexuality and anxiety, or of the parasympathetic (vagus) and sympathetic. They share a common origin; they form a unit; they are a function belonging to the same vegetative system, but the vegetative functional *unity* divides into *two antithetical functions and directions of current. Anxiety becomes the opposite of sexual excitation; it corresponds to a direction of electrical energy function and flow opposite to that of sexuality.* Similarly, the vagus and sympathetic are opposed to each other. The following diagram illustrates this:

The unity and antithesis of the autonomic nervous system.
Vagus = pleasure; Sympathetic = anxiety

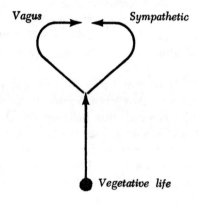

We do not reflect on nature in dialectical terms. The natural process is itself dialectical. Far from being mere conceptualizations, the differentiation of the homogeneous whole and the antithetical relationships can be seen, measured, and photographed in reality. Through its vegetative apparatus, each living organism is part of the whole of living nature, the result of the general splitting of the one vegetative life into billions of kinds of life. And each living organism at the same time contrasts with the rest of the vegetative world as well as with all other living organisms, whether in sexual intercourse, in the act of eating, or, ultimately, in the passive role of the object eaten. The following diagram illustrates the process of vegetative differentiation and antithesis:

Vegetative differentiation and antithesis in organic life

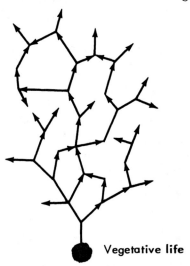

Vegetative life

Since Darwin, there has been no doubt about the reality of this fragmenting of the phylum into different species. The *dialectical law of development (fragmenting and opposition and unification)* is *the basic law of all organic development*. The branching of the trunk of a tree into boughs, then twigs, and finally into leaf stalks, of these leaf stalks into leaves, of the leaves into a central midrib, from which again fine veins branch out (primary, secondary, tertiary, etc., venation), is dialectics in the reality of nature.

The following figure (58) shows the division and opposition in the structure of leaf arrangement.

The branching of the nervous system and of the system of blood vessels in the body follows the same law. As the center of the circulatory system, the heart divides into arteries and veins, and these in turn branch continuously into an infinitely large number of subdivisions, until, finally, the very fine capillaries, which carry blood in opposite directions, come together again. In other words, they contrast with each other and then are united.

Figure 58. Genetic dissociation in the structure of a foliage plant. Dialectical materialism is not just a philosophical method but also truly reflects the processes occurring in nature.

This basic idea cannot at present be applied with absolute certainty to the nervous system, because the functional relationship between nerve ending and muscle fiber has still not been completely explained. If Kraus is right in stating that the nervous system represents a syncytium, if the muscle and the nerve form a functional unity, then it would be easier to apply the dialectical law of development to the nervous system as well. Many a dispute—such as whether the stimulus-conducting apparatus of the heart is nervous or muscular in nature—would thus be resolved.

The essential factor for understanding the dialectical function of life is that this splitting of the whole does not affect its homogeneity because, no matter how fragmented and antithetical the vegetative nervous system may be, it nevertheless functions as a *unitary indivisible* whole.

Let us now try to link this principle more closely to the actual subject of the present discussion. The important question is how metazoal function is related to protozoal function. The metazoan—for example, the human body—was thought of as being constructed of individual cells, with certain groups of individual cells performing certain individual functions such as bile secretion, excretion of urine, formation of blood corpuscles, innervation, etc. Human and animal bodies were seen as being constructed along the lines of a well-functioning administration. At the "top" is the brain and the process of inner secretion which direct functions, while the individual organs are assigned various "tasks." "Eating" and "procreation" were thought of as functions of the total organism, but always in the background was the idea that this organism contained a certain something that wanted to preserve the individuals and the species—a thoroughly metaphysical principle. The same functions were also ascribed to the protozoa. We know, however, that the metazoal organism functions, as a whole, just as homogeneously as does the protozoan. Like the amoeba, when a metazoan stretches, it does so as an entire organism. When frightened, it retracts as an entire organism, in the same way that the amoeba retracts all its pseudopodia. Thus, all living organisms are permeated by a basic com-

## The unity of living functions

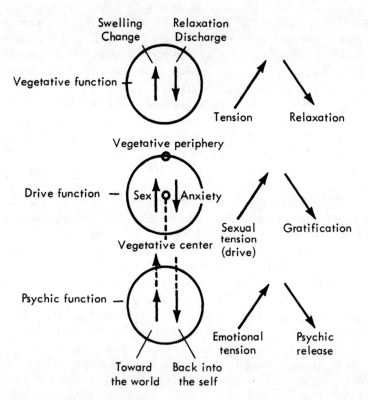

monality that follows a given set of laws, regardless of all the differentiation and antithetical relationships.

It was at first surprising, indeed almost staggering, to note that the antithesis of psychic pleasure and anxiety is functionally identical with the physiological vagus and sympathetic functions, that the vagus and sympathetic functions are identical with the organic substances lecithin and cholesterin, and that these, in turn, are identical with the effect of inorganic substances such as potassium and calcium. If we reverse the sequence and proceed from potassium or calcium via lecithin or cholesterin to vagus or sympathetic and finally to pleasure or anxiety, we see how a function develops to higher, complicated, differentiated forms. The fundamental character of life governs all development and all products of development. It consists of the two antithetical functions of expansion and contraction, tension and relaxation, charge and discharge.

## SOME REMARKS ON BIOGENESIS

Now that we have familiarized ourselves with the general principles of dialectical-materialistic and functional thought, we can dare to test its special application to the evolution of living organisms. The following remarks offer only provisional guidance.

The inorganic and organic spheres are not two strictly distinct, totally unconnected areas with no bridge between them. Since, in principle, organic life exhibits the same physical and chemical processes as inorganic matter—in particular, mechanical and electrical processes—inorganic and organic matter are, broadly speaking, functionally identical. To this extent, our methodological view coincides with that of the materialistic mechanist. Our view differs from that of the latter, however, because in addition to this functional identity we must simultaneously assume a functional antithesis. The point is not to assume, in abstract epistemological terms, a functional antithesis in addition to a functional identity, but actually to show in a practical manner that such an antithetical relationship exists and to reveal what concrete form it takes. Hartmann writes in the section on "Begriff und Umfang der allgemeinen Biologie" in his main work *Allgemeine Biologie* (3rd ed.):

> It is in fact not possible at the present time to determine and define life as having a specific chemical and material structure in the same way that one can define certain natural inorganic bodies such as minerals and crystals; instead, living organisms additionally embody events and processes of a special kind which give the relevant structures and systems their characteristic stamp as organisms; and when such processes and events are lacking or cease to exist, it is no longer possible to talk of life.
>
> Chemical, mineralogical, etc., bodies are systems in a steady state of equilibrium, a true chemical equilibrium. Admittedly, living systems are also bodies which give the impression of being more or less durable. And yet what a difference there is! In the latter case, the equilibria are anything but stable, and, instead, the permanency of the individual is, from the physico-chemical standpoint, an illusion. It is only achieved by a process of continuous change, a continuous buildup and breakdown of the chemical substances contained in the systems, and a continuous fluctuation of energetic forces; these are, in fact, only *dynamic* equilibria. A continuous flow of material and energy metabolism takes place in living systems and maintains them in an apparent state of stable equilibrium.

Hartmann stresses that all processes that have proved to be characteristic of life consist of metabolism, stimulus phenomena, and changes in shape. The problem now is to decide what gives rise to these fundamental characteristics of living matter; in particular, to stimulus phenomena and the change of shape. If we regard them from the start as, so to speak, the explanatory principle, the question of their nature and mechanical functioning, and how these differ from the inorganic sphere, still remains unanswered.

We understand that the organic and inorganic spheres are

functionally identical in mechanical and electrical terms. So far, our experiments have yielded the following results:

1. *The fundamental mechanical and electrical factors that predominate in inorganic matter come together in living organisms in a functional relationship that is specific to life.* There is no inorganic process in which mechanical filling (swelling) would turn into electrical charge, then into electrical discharge, and then into relaxation. The given *sequence* of the mechanical and the electrical functions is specific to life and *differentiates it from the inorganic sphere.*
2. The tension → charge → discharge → relaxation process not only distinguishes life from lifeless matter, but, if the preceding statement is correct, it must in principle govern all living functions. Therefore, it should be possible to show that all vegetative functions follow the given rhythm of mechanical tension → charge → discharge → relaxation. This formula, which we might call the *life formula,* was discovered in the function of the orgasm. One could just as well, although not so easily, have determined the formula in the automatic movement of the intestine, the heart, or in cell division, etc.
3. The life (or orgasm) formula governs all vegetative functioning, not only in the biological organism as a whole but in each of its parts. Each metazoal cell is subject to the law of life that we have summarized in the tension-charge formula, both as a distinct part and as a component of the total organism. It should thus be possible to demonstrate the tension-charge function both in the uniform overall function of any living organism and in the detailed functioning of its individual cells. This holds true provided that one does not examine the mechanical processes separately from the chemical-electrical processes, and vice versa, but, instead, always considers them in relation to the life rhythm.

Life, as a fundamentally homogeneous function, emerges from the inorganic. But, in the realm of life, voluntary action develops out of vegetative unconscious functioning. This applies to the most sophisticated product of development *at present;* i.e., *consciousness,* with all its functions. We are of course still a long way from even beginning to understand all the differences within the sphere of vegetative life. We already have a general view of part of the fundamental commonality, but we do not yet know all the circumstances and conditions which, within the sphere of life, cause a new function to arise from the old.

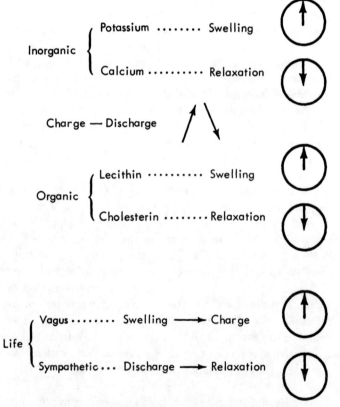

The unity of living matter and non-living matter

## Dialectical-Materialistic Method of Thinking and Investigation 175

So much for the fundamentals. The following assumptions can be based on the foregoing views:

1. Life cannot have formed as a result of "spontaneous generation" in the sense that at some time, at one single place in the universe, somehow or other life was generated and then spread without interruption. The theory of cosmic creation, of living substances coming down from space onto the earth, is unproven and also improbable, because it is too complicated. One would at least have to show that it is probable that the conditions for such a transfer of life from space are present. Protein floccules or spores from space would have had to survive all the obstacles and dangers of such a perilous journey and have landed simultaneously *all over the globe*.
2. Direct observation of living processes, particularly of vegetation, forces us to assume that life developed from inorganic matter under extremely *simple* and *natural* conditions.
3. These natural and simple conditions are present *everywhere today*. Life requires certain substances with certain properties; namely, carbon, oxygen, hydrogen, and nitrogen. These substances are found *everywhere*. For them to produce living functions, certain conditions have to be met which permit the leap from mechanical filling (swelling) to electrical charge. These conditions, both the swelling and the electrical processes, exist everywhere.
4. Life can be generated at any time and at any place where the necessary substances and conditions exist. Nature does not possess any weighing scales or electrical instruments. There must therefore be a principle that determines the choice of the quality and quantity of the necessary conditions. This would be the principle of *self-regulation* in the creation of life. Thus, new life is generated everywhere by the hour. It is only necessary to see it in its overall context.
5. Anything that promotes swelling and charge simultaneously promotes life.
6. Continuously originating life has to be differentiated. Here

again, we have to distinguish between those forms of life that develop from decomposing organisms and those forms that develop through the organizing of inorganic matter. Not only does life originate, but the originated forms are *propagated* and repropagated.

7. If life forms from lifeless matter and then becomes lifeless matter again, there is a cycle between inorganic and living matter as well as within living matter. When it dies, the multicellular organism disintegrates into unicellular organisms and into inorganic matter. The unicellular organism forms again from both sources.
8. In principle, death can be seen as the cessation of a function at one of the three main points in the life rhythm:
    a. Shrinking, or loss of swelling substance (mainly water), renders impossible the first act of biological functioning; namely, mechanical tension (dying of thirst).
    b. The transition from swelling to charge is disturbed. This interrupts the automatic sequence of the life function (death through heart attack).
    c. The transition from discharge to relaxation could be disturbed.
    d. Finally, the transition from relaxation to renewed swelling may be disturbed.

When ether-anesthetized experimental mice are observed dying, it is noted that there are phases in the death process. First, the mouse defends itself against the poison by increasing its motor activity. Then it collapses. The respiration, which is at first very rapid, becomes irregular and finally stops. However, death has not yet occurred. After cessation of respiration, the entire body convulses, due to discharges of the electrical energy of the body. No new buildup of charge occurs, since the respiration—and thus the combustion—no longer functions. The convulsive movements die away and sometimes the heart still continues to beat for a short while. Thus, one function after the other is extinguished. Depending on the importance of each function for the life process, it may or may not be possible to resuscitate the animal.

One day, the means will be found to restore the functions, particularly swelling and charge. But today such hopes lie in the realm of scientific daydreams. Yet, why shouldn't we dream? Today's dream is often tomorrow's reality.

# *Appendix*

## PRODUCTION OF BIONS FROM STERILIZED BLOOD CHARCOAL (CARBO SANGUINIS)

1. Powdered blood charcoal is sterilized for two hours at 190°C in a small dish in the dry sterilizer. Equal quantities of beef broth and 0.1N KCl are mixed and autoclaved for half an hour in two test tubes at 120°C. Two test tubes containing pure 0.1N KCl are similarly treated.
2. With a metal spatula, a small quantity of sterilized blood charcoal is *heated to incandescence* in the *upper part* of the benzene gas flame. This substance is then *dry*-inoculated onto blood agar as a control of the sterility. No growth should occur.
3. Another pinch of blood charcoal is heated in similar fashion on a spatula tip and divided among each of the four test tubes. These are then shaken so that the blood charcoal is uniformly distributed in the liquid. They are then put in the incubator. Very soon the blackish color of the colloid gives way to a gray and slightly cloudy appearance. After twenty-four to forty-eight hours, a fine cloudy gray turbidity is present. It can be detected by gently shaking the test tube.
4. A drop is taken under sterile conditions from each of the KCl and broth + KCl preparations and studied at a minimum magnification of 3000× under a binocular microscope (inclined focusing tubes). The bions must exhibit vigorous movement and possess a large number of motile and contractile clusters of vesicles. A test of their electrical characteristics should show that they are strongly positively charged ($=$ mi-

gration toward the cathode). If so, *fresh* egg medium is generously inoculated and placed *horizontally* in the incubator. After twenty-four to forty-eight hours, most of the inoculated egg nutrient media are covered more or less densely with small gray, round hummocks. If these hummocks are numerous enough, they are then spread out on the same medium, using a heated platinum wire, and the cultures are again placed in the incubator.

5. After a further twenty-four hours or slightly longer, a dense light-gray, *bluish shimmering*, coherent growth occurs. This is transferred to clear fresh blood agar nutrient medium. Within twelve to twenty-four hours a dense, creamy, *blue-gray* growth occurs. When viewed under the microscope, it is seen to consist of a pure culture of round and ovoid cocci and a large number of vigorous, rapidly moving contractile clusters of vesicles. Their electrical charge is positive.
6. These *blood charcoal bions* are injected subcutaneously into the backs of mice. They should not produce any pathological reaction.
7. Repeated autoclaving of the cultures yields further cultures on egg nutrient media.

## THREE SERIES OF EXPERIMENTS BASED ON THE TENSION-CHARGE PRINCIPLE
### by Roger du Teil

The first two series of experiments which are reported below correspond in principle, but with modifications, to the procedure worked out by Dr. Wilhelm Reich of Oslo, in keeping with his overall biological theory.

The third series of experiments, designed and carried out by me, follows the same basic pattern, but the method, results, and also the conclusions that can be drawn from them deviate distinctly from the other experiments. This last series of experiments confirms, simplifies, and concentrates Reich's method.

Dr. Reich's synthetic theory essentially consists of equating the psychic tendencies *"toward the world"* and *"into the self, away from the world"* and the accompanying feelings of *pleasure* and *anxiety* with, on the one hand, the *"sympathetic"* and *"parasympathetic"* nervous systems which control the shifts in fluids in the organism that accompany these feelings; and, on the other hand, with the proteins and other chemical substances which promote these movements in the higher organisms and which in the lower organisms replace the nervous system of the metazoa. The psychic tendencies are also equated with the electrical charge and discharge processes which accompany the shifts in fluids and which, through the interplay of *expansion* and *contraction*, while continuously maintaining the fundamental antithesis between *"center"* and *"periphery,"* constitute life itself.

We thus have the pairs of antagonists *lecithin-cholesterin* on the one hand and *potassium-calcium* on the other. When introduced into a colloidal liquid where the movement arises from and is increased by Brownian movement of fine coal particles and where a large number of limiting membranes form, these antagonists cause organisms to appear in the liquid which have all the properties of life and which are culturable.

The three series of experiments described below were carried out by me after my return from Oslo. Dr. Wilhelm Reich, whose work I have been following and verifying at his request for several years, explained his own control procedures to me in his laboratory in Oslo and invited me to verify and compare them with my own procedures. The apparatus which I have constructed and which are described here were designed with the aim of simplifying the control experiments and making them easier and more reliable to apply.

## FIRST SERIES
### "Bion" Experiment

*First Experiment*

An apparatus is constructed along the following lines: Two containers are connected with one another in such a way that each can be sealed off from the other and placed in an autoclave. A third glass container (test tube with broth) forms a further part of the apparatus, which should thus be a strictly closed system (see Illustration 1).

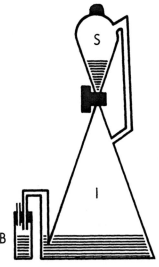

Illustration 1. Sy-Clos

The apparatus consists essentially of two containers, one arranged on top of the other, which can be cut off from each other by a glass stopcock. A narrow-gauge tube links the two upper sections so that air from the lower container can pass into the upper container when the liquid from the upper vessel runs into the lower. A double right-angled capillary outlet tube is attached to the lower container, and to it is connected, by a sealed rubber stopper, the test tube containing the broth to be inoculated. The two original mixtures can thus be placed separately and yet together into the autoclave, and with them also the broth which is intended to demonstrate whether the organisms obtained are living. After removal of the apparatus from the autoclave, the two solutions can be mixed with each other by opening the stopcock. When the appointed time is up, it is sufficient to heat the upper container, so that the expanding air drives a few drops of the mixture to be tested into the capillary tube connected with the broth. Thus, once the apparatus has been taken from the autoclave, the whole process takes place in a strictly closed system.

This apparatus, which for ease of identification we will call "*Sy-Clos*," thus excludes any possible objection that the mixture may be accidentally infected by an airborne germ or by a microorganism introduced at some time during the experiment. Furthermore, if Reich's original procedure is followed, preparation 6 would have to be placed once more in the prepared condition in the autoclave in order to kill any microorganisms that may have been introduced during the experiment. This sterilization procedure also carries the risk that the bions, too, might be killed off, or at least their vitality might be weakened, damaging their culturability. In the apparatus of the Sy-Clos series, no exogenous infection is possible following sterilization prior to the mixing; therefore, resterilization is not necessary and the bions can be cultured in exactly the way that they form.

It should be noted here that, in those cases where it is not necessary to use a strictly closed apparatus to determine the purely endogenous origin of the bions, the presence of the bions in the final mixture can be proven by carrying out microscopic examination of a sample taken *immediately* after this mixture is prepared. In this way, the hypothesis stating that spores developed which were already contained in one of the substances used can be excluded.

Furthermore, when this examination is carried out for a long time in the concavity of the slide, the heat from the microscope lamp causes the microorganisms to multiply to such an extent that the preparation becomes densely populated and within the space of fifteen minutes it is no longer suitable for observation purposes.

On September 3, 1937, the two containers were filled, one after the other, as follows.

Lower container: equal parts of 0.1N KCl and Ringer's solution, several milligrams of red gelatin dissolved in 3 cc 0.1 N KCl, two crystals of cholesterin, ½ centigram of finely pulverized coal (coke) heated to incandescence on a spatula in a Bunsen flame, a few drops of egg white, a few drops of egg yolk, a few drops of milk, all taken under sterile conditions. Then the stopcock was closed.

Upper container: about 10 milligrams of lecithin were pulverized in a mortar and added to 1 cc of 0.1N KCl. When the mixture was turbid and had assumed a yellowish color, it was poured into the upper container of the apparatus.

Sterile broth was put in the tube connected to the side of the lower container. Once the upper stopper had been replaced by a plug of cotton wool, which allowed air to escape during the sterilization process, the entire apparatus was placed in the autoclave.

The first sterilization was carried out on September 4 for three-quarters of an hour at 134°C.

A second sterilization was carried out on September 5, twenty-four hours later. Same temperature, one-hour duration.

A third sterilization was carried out on September 6, twenty-four hours after the second: 130°C, one half hour.

After autoclaving, the plug of cotton wool was immediately replaced by the sterilized stopper, which was sealed in place with paraffin. Then the stopcock was opened and the solution from the upper container was mixed with that in the lower. The stopcock was closed and also sealed, as was the stopper and all joints in the broth tube at the side of the apparatus.

On September 10, four days later, the upper container was

h-tube

Figure 59. Sy-Clos apparatus for preparation 6, du Teil system

heated and a few drops were inoculated into the broth in the test tube. Because of its size, the apparatus could not be fitted completely into the incubator. Twenty-four hours later, the broth still showed no signs of turbidity. A new inoculation was carried out with a few drops by heating the upper container and expanding the air. Twenty-four hours later, on September 12, the broth exhibited very clear turbidity and further inoculations were made from it onto agar and broth.

From September 13 on, this last broth exhibited a very clear turbidity and there was a gray, creamy culture on the agar which initially consisted of small round colonies but then rapidly spread over the entire surface.

These two cultures made on September 12 were inoculated again into broth and onto agar on September 13. These four tubes yielded cultures of the same type. The inoculations and cultures have since then proceeded normally.

Microscopic examination of the first mixture, as well as of the cultures, revealed very large quantities of round or ovoid, motile organisms which, as they multiplied, more frequently formed chains than clusters. These are the "bions" first produced and named by Dr. Reich in Oslo.

My own investigations have shown that the bions are gram-positive. Furthermore, I was able to detect a sensitivity toward certain poisons. They can be killed—that is, they can be robbed of their motility—by bleaching liquid (potassium hypochlorite) in about fifteen minutes and by alcohol in about one minute. Ether dissolves them within a few minutes.

*Second Experiment*

A similar experiment was carried out on September 15. The same apparatus and the same procedures were used, but *no milk was added to the mixture.* Bions were formed as in the first experiment. In this case, however, they were inoculated sooner, on September 18. Here again, the entire process took place in a closed system and the broth immediately became very turbid. When examined under the microscope the broth culture proved to be much more lively than those in the first experiment, although they were fewer in number.

## Third Experiment

In this experiment, a new piece of apparatus was used, built along the same lines as the first but simpler in design. It consists of four adjacent tubes; three of them lie in one plane, and the fourth is arranged at right angles to them. The tubes are connected in such a way that the two bion mixtures can be combined and the inoculation into broth can be carried out simply by tilting the apparatus. Finally, one tube serves as a control; the broth contained in it is exposed to the same effects but is not inoculated. This apparatus is called, for the sake of convenience, a Sy-Clos tube (see Illustration 2).

On September 18, very small quantities of bions were produced in the Sy-Clos tube. No milk was added to the mixture. Since the gelatin used was colorless, a trace of methylene blue was added, so that the mixture took on a color that indicated its gelatin content. The whole system was autoclaved at 130°C for one hour (which corresponds to three successive sterilizations), and mixing was carried out by tilting the apparatus. The mixture took on a greenish opalescent color reminiscent of absinthe. The inoculation was carried out on the twentieth. On the twenty-first, as in the previous experiments, a very pronounced, homogeneous, and moiré-patterned turbidity was present. On the twenty-second, heat was applied to the apparatus, thereby driving a few drops through the capillary tube and thus inoculating an agar nutrient medium and a gelatin–egg white nutrient medium with the broth. Cultures resulted from both inoculations.

When this broth was examined under a microscope, a large number of ovoid bions linked in chains, as well as some rigid lecithin tubes, inside which cocci were in the process of forming, were observed. The presence of these lecithin tubes was due to the fact that a rather large amount of material was inoculated and these organisms were entrained as a part of the mixture.

Fuchsin-stained microscopic preparations of this particularly characteristic broth were put aside.

Illustration 2. Sy-Clos tube

a. Mixture I
b. Mixture II
c. Broth to be inoculated
d. Control broth °

° *Instructions for using the Sy-Clos tube*

Put the lecithin mixture in tube *a*.

Put the KCl-Ringer's-coal, etc., mixture in tube *b*. (Use very small quantities.)

Put the broth into tube *d* first.

Insert the stopper into tube *d* so that it closes the connection with tube *c*; this prevents premature inoculation occurring during autoclaving or during the first manipulations. The entire system is then autoclaved.

Next, the contents of *a* and *b* are mixed in *a* or *b*. (It is best to do this in *a*.)

To bring about inoculation, a drop of the ready mixture is introduced into tube *c*, which has so far remained empty, and the stopper in *d* is slightly withdrawn so that it uncovers the opening of the connecting tube and permits the broth to transfer from *d* to *c*. A small quantity of broth remains behind in *d* for control purposes.

Now a plug of cotton wool can be used in *d* to prevent infection of the control broth by the air in the other tubes.

## SECOND SERIES
### "Incandescent Coal" Experiment

*First Experiment*

On August 19, the following mixture was prepared in a test tube: about 10 cc of 0.1 N KCl, a drop of sterile gelatinized blood, about 1 centigram of finely pulverized coke heated to incandescence. (The remainder of the gelatinized blood is still today, on September 23, completely sterile.)

On August 28, a broth was inoculated from this mixture. The *result was positive,* and uniform turbidity was achieved.

On August 29, an inoculation was carried out from the broth onto gelatin. A *positive result* was achieved. A gray culture similar to that of the bions was obtained.

On the same day, August 29, the tube containing the stock mixture was boiled twice at 100°C, for a half hour each time, in ambient air. Then inoculations were made from this into broth. *The result was positive.*

On September 3, the tube containing the stock solution was autoclaved for a half hour at 134°C. Then an inoculation was made into broth; a positive result was obtained. On September 4, inoculations were made from the broth into gelatin—with *positive results.*

On September 5, the tube containing the stock solution was autoclaved again for three-quarters of an hour at 130°C and then inoculations were made from it into broth. The *result was positive.* In order to avoid any infection, the inoculation was carried out in the following way: The sterile liquid was placed in a sterile dropper which was positioned as a stopper in the tube containing the broth to be inoculated. The actual inoculation took place automatically as a result of the increasing pressure in the autoclave. It was not touched in any way, air was completely excluded and the temperature was set at a level which is accepted as lethal for all microorganisms.

On September 6, inoculations were carried out from this broth onto gelatin. The *result was positive.*

*Second and Third Experiments*

On September 27 and 28, the same experiment was repeated twice, this time, however, replacing the gelatinized blood with sterile broth and varying the proportions of the broth and KCl each time.

Like the first experiment, these two experiments also yielded positive results, despite all attempts to sterilize the tubes containing the stock solution.

## THIRD SERIES
### Experiment with Pure Potassium Chloride (KCl)

If Pasteur's hypothesis is correct, then of the three substances used in the preceding series of experiments, neither the coal, which was heated to vigorous red, even white, heat, nor the broth, whose sterility was guaranteed by the clarity of the broth in the control test tubes, can be suspected of being non-sterile. The only remaining possibility was to suspect the KCl, but this was obtained in pure form, in crystals, from the firm of Poulenc, and it was dissolved by boiling continuously for an hour and a half at 100°C.

This third series of experiments was undertaken for the purpose of clarifying the role of the KCl in the preceding experiments.

On August 31, a flask containing this *triple-boiled* solution was placed in the autoclave and sterilized in the customary manner. A broth was then immediately inoculated from the solution. A *positive result* was obtained. An inoculation made from this broth on September 2 onto agar *yielded a very strong culture*. On September 3, the same flask containing KCl (a dropper flask that permits automatic inoculation) was autoclaved once more for a quarter of an hour at 130°C. Inoculation was made into broth. A *positive result* was achieved.

On September 5, the same flask was again placed in the autoclave. At the same time, KCl from this flask was placed in a sterilized dropper which sat on top of the tube containing the broth. The inoculation was brought about by the steam pressure inside the autoclave—i.e., air was excluded—and at a temperature of 130°C. *A positive result* was achieved. This broth was inoculated onto agar on September 8 and gave *a very good gray culture,* which admittedly took longer to develop than the previous cultures.

On September 8, the KCl in the same flask was reautoclaved and directly inoculated onto agar. After four days, a *positive result* was achieved in the form of a light gray, almost white culture. Further inoculation onto another gelatin on September 11 again yielded a *positive result;* namely, a very distinct whitish-gray culture.

After this, I made two "h-tubes," as the apparatus is called, which consist essentially of two test tubes, the opening of one tube being fused into the wall of the other at about midpoint. The lower half of the test tube then runs parallel to the other straight test tube. The apparatus resembles in appearance a small *h.* The single opening, which is located in the upper section, is sealed with a rubber stopper or can be fused shut in a flame; either way, it can be hermetically sealed. A fine drawn-out tube, which is attached to the side of the straight tube and fused shut in a flame, permits later sampling of the untouched liquid (see Illustration 3).

On September 8, KCl was taken from the flask, *which had now been sterilized for the fifth time,* and placed in the straight section of the apparatus. Sterile broth was placed in the angled section. The entire apparatus was put in an autoclave at 130°C for one hour. When the apparatus was removed from the autoclave, it was simply tilted so that the broth flowed over and came into contact with the small amount of KCl in the bottom end of the other tube. The broth was thus inoculated. At the same time, some broth remained behind as a control in the angled section of tube.

This experiment, which was carried out under unimpeachable conditions of sterility as regards both the substances and the apparatus, *yielded a positive result.* From the second day on,

## Illustration 3. h-tube

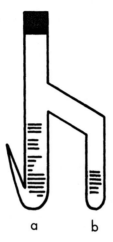

a. 1st phase: KCl
   2nd phase: KCl + broth
b. 1st phase: Broth
   2nd phase: Control broth

the broth in the straight tube, compared with the control, exhibited an admittedly very weak but nonetheless perceptible turbidity. A further inoculation onto gelatin, which was carried out on the fourteenth, yielded after a few days an initially very weak culture which possessed no chromogenic properties; that is, it was just as transparent as the nutrient medium. However, after a few more days, on those spots where the drops of broth were placed, a gray culture formed which seemed to be favorably influenced by a lower temperature than that in the incubator. Two microscopic samples taken from the weak and the creamy portion of the culture revealed that the organisms that had developed were completely identical. The tendency toward chain formation was more strongly in evidence in the later organisms.

On September 9, an identical experiment was carried out with the second h-tube, and again a *positive result* was achieved.

While the turbidity of the original broth was scarcely detectable, an inoculation onto gelatin carried out eleven days later quickly yielded a transparent culture similar to the first, but thick and strong. When examined under the microscope, it, too, contained the same organisms.

It still remains to be mentioned that several experiments in which KCl, heated to incandescence, was added to the broth also yielded very positive results. On the other hand, an experiment involving KCl that had fused during heating yielded only a doubtful result.

*Conclusions*

This last series of tests permits only two interpretations. *Interpretation in accordance with Pasteur's views:* KCl must contain germs which cannot be killed even by repeated application of usual sterilization procedures. It would then remain to be explained just how these germs can be invisible in the pure KCl and then appear at the precise moment when the KCl is mixed. At any event, the discovery of these germs is a matter of some interest.

*Interpretation in accordance with Reich's theory of synthesis:* The broth contains all the substances which we used for the mixtures in the bion experiments, in particular the antagonists lecithin-cholesterin and potassium-calcium. It is thus conceivable that an excess of potassium in the broth can directly achieve the same effect that was observed in the first mixture, either because this enables the substances to organize or because the vitality which these substances had lost as a result of sterilization is restored to them. Since the sterility of the KCl can only be tested by introducing it into the broth, there is no way of deciding which of these two views is correct.

Nice, September 29, 1937

# Bibliography

Abel, R. Überblick über die geschichtliche Entwicklung der Lehre von der Infektion, Immunität und Prophylaxe," *Handbuch der pathogenen Mikroorganismen,* 3rd ed., edited by W. Kolle, R. Kraus, P. Uhlenhuth. Vol. 1, pp. 1–31. Jena and Berlin/Vienna, 1929.

"Abiogenesis," *Encyclopaedia Britannica,* 14th ed., Vol. 1.

Bastian, Henry Charlton. *Modes of Origin of Lowest Organisms.* London, 1871.

———. *The Beginnings of Life.* London, 1872.

———. *Evolution and the Origin of Life.* London, 1874.

Bavendamm, Werner. "Die Physiologie der schwefelspeichernden und schwefelfreien Purpurbakterien," *Ergebnisse der Biologie,* Vol. 13, pp. 1–53. Berlin, 1936.

Bavink, Bernhard. *Ergebnisse und Probleme der Naturwissenschaften,* 4th ed. Leipzig, 1930.

Bertalanffy, Ludwig v. "Studien über theoretische Biologie, I und II," *Biologisches Zentralblatt,* Vol. 47. 1927.

———. *Theoretische Biologie I.* Berlin, 1932.

Brauner, L. *Die Pflanze.* Berlin, 1930.

Brauns, R. "Flüssige Kristalle und Lebewesen," *Centralblatt für Mineralogie, Geologie und Paläontologie,* section A: *Mineralogie und Petrographie,* p. 256. Stuttgart, 1931.

———. Discussion of F. Rinne's "Grenzfragen des Lebens," *Cbl. f. Min.,* section A, p. 166. 1931.

———. *Neues Jahrbuch für Mineralogie usw.,* 1931.

    (a) Discussion of a paper by F. Rinne, "Spermien als lebende flüssige Kristalle" (*Naturwissenschaften,* Vol. 18, No. 40, p. 837, 1930), pp. 1–2.

    (b) Collective report on the discussion on "Flüssige Kristalle" in *Zeitschrift für Kristallographie,* Vol. 79, Nos. 1–4, section A, pp. 322–31. Leipzig, 1931.

———. "Flüssige Kristalle," *Handwörterbuch der Naturwissenschaften,* 2nd ed., Vol. V, pp. 1159–70. 1934.

## BIBLIOGRAPHY

Brüel, L. "Zelle und Zellteilung, zoologisch," *HWB d. Nat.-Wiss.*, 2nd ed., Vol. X. 1935.

Cohn, Ferdinand. "Untersuchungen über Bakterien I," *Beiträge zur Biologie der Pflanzen*, Vol. I, No. 2. Breslau, 1872.

———. "Untersuchungen über Bakterien II," *Beiträge zur Biologie der Pflanzen*, Vol. I, No. 3. 1875.

———. "Untersuchungen über Bakterien IV: Beiträge zur Biologie der Bacillen," *Beiträge zur Biologie der Pflanzen*, Vol. II, No. 2. 1876.

Cohn and Mendelsohn, B. "Untersuchungen über Bakterien IX: Über die Einwirkung des elektr. Stromes auf die Vermehrung v. Bakterien," *Beiträge zur Biologie der Pflanzen*, Vol. III, No. 1. 1879.

Cohn and Miflet. "Untersuchungen über Bakterien VIII: Untersuchungen über die in der Luft suspendierten Bakterien," *Beiträge zur Biologie der Pflanzen*, Vol. III, No. 1. 1879.

Dastre, A. *La vie et la mort*. Paris. 1916.

Dembowski, J. Report on A. I. Oparin's *Der Ursprung des Lebens auf der Erde* (Moscow/Leningrad, 1936), in *Berichte über die wissenschaftl. Biologie*, Vol. 41, No. 3/4, pp. 145–46. 1937.

Driesch, Hans. "Das Wesen des Organismus," *Das Lebensproblem im Lichte der modernen Forschung*, edited by Hans Driesch and Heinz Woltereck, pp. 384–450. Leipzig, 1931.

Ehrenberg, Rudolf. *Theoretische Biologie, vom Standpunkt der Irreversibilität des elementaren Lebensvorganges*. Berlin, 1923.

Fechner, G. T. *Einige Ideen zur Schöpfungs- und Entwicklungsgeschichte der Organismen*. Leipzig, 1873.

Fetscher. Report on Glässer's *Zelle, Bakterium und mikroskopisch unsichtbare Lebewesen*, in *Berichte über die gesamte Physiologie u. exper. Pharmakologie*, Vol. 98, No. 3/4, p. 200. 1937.

Frank, Philipp. "Das Kausalgesetz und seine Grenzen," *Schriften zur wissenschaftlichen Weltauffassung*, Vol. 6. Vienna, 1932.

Franz, Victor. "Probiologie und Organisationsstufen," *Abhandlungen zur theor. Biologie*, No. 6, edited by J. Schaxel. Berlin, 1920.

Gause, G. F. "Raumaufbau des Protoplasmas," *Erg. d. Biol. XIII*, pp. 54–92. Berlin, 1936.

Gotschlich, E. "Allgemeine Morphologie und Biologie der pathogenen Mikroorganismen," *Hdb. d. path. Mikroorg.*, Vol. I, No. 1, pp. 33ff. 1929.

Haeckel, Ernst. *Anthropogenie oder Entwicklungsgeschichte des Menschen*. Leipzig, 1877.

Hartmann, Max. *Philosophie der Naturwissenschaften.* Berlin, 1937.
Huizinga, D. "Zur Abiogenesisfrage," Pflueger's *Archiv f. d. ges. Physiologie,* Vol. 7, pp. 549 ff. 1873.
——. "Weiteres zur Abiogenesisfrage," Pflueger's *Archiv,* Vol. 8, pp. 180 ff. 1874.
——. "Zur Abiogenesisfrage (III)," Pflueger's *Archiv,* Vol. 8, pp. 551 ff. 1874.
——. "Zur Abiogenesisfrage (IV)," Pflueger's *Archiv,* Vol. 9, pp. 62 ff. 1875.
Gerhardt, U. "Experimentelle Urzeugung?," *Medizinische Klinik,* No. 2, p. 6. 1906.
Gscheidlen, Richard. "Über die Abiogenesis Huizingas," Pflueger's *Archiv,* Vol. 9, pp. 163 ff. 1874.
Jaworski, Hélan. *Le Géon ou la terre vivante.* Paris, 1928.
Jensen, Paul. "Leben," *HWB d. Nat.-Wiss.,* Vol. X. 1931.
——. "Kausalität, Biologie und Psychologie," *Erkenntnis,* edited by Carnap and Reichenbach, Vol. 4, pp. 165–214. Leipzig, 1934.
Kammerer, Paul. *Allgemeine Biologie.* Stuttgart/Berlin, 1920.
Koch, Robert. "Die Aetiologie der Milzbrandkrankheit," *Beiträge zur Biologie der Pflanzen* (Cohn), Vol. II, No. 3. pp. 399 ff. 1877.
"Krankheitserreger und Wachstumserreger." Bericht von der Versammlung Deutscher Naturforscher und Ärzte zu Dresden, 1936. *Frankfurter Zeitung,* Sept. 25, 1936.
Le Dantec, Felix. *La lutte universelle.* Paris, 1920.
Leduc, Stephane. *Die synthetische Biologie.* Halle, 1914.
Lehmann, Otto. "Fliessende Kristalle und Organismen," Roux's *Archiv für Entwicklungsmechanik,* Vol. 21, pp. 596–609. 1906.
——. *Flüssige Kristalle und die Theorien des Lebens.* Leipzig, 1906.
——. *Die neue Welt der flüssigen Kristalle.* Leipzig, 1911.
Lieske, R. "Bakterien und Strahlenpilze," *Handbuch der Pflanzenanatomie,* section II, part 1, Vol. 6. Berlin, 1922.
Lippmann, E. O. v. *Urzeugung und Lebenskraft.* Berlin, 1933.
Loeb, Jacques. *Vorlesungen über die Dynamik der Lebenserscheinungen.* Leipzig, 1906.
——. *The Mechanistic Conception of Life.* Chicago, 1912; Harvard University Press, 1964.
Mach, Ernst. *Erkenntnis und Irrtum.* Leipzig, 1905.
Mary, Albert. "La vie merveilleuse des minéraux," in H. Jaworski's *Le Géon ou la terre vivante.* Paris, 1928.

Miehe, H. "Sind ultramikroskopische Organismen in der Natur verbreitet?," *Biol. Zbl.*, Vol. 43, pp. 1–15. 1923.

Moigno, Abbé. *Introduction to Tyndall-Pasteur's Les microbes organisés.* Paris, 1878.

Muenden, Max. "Der Chtonoblast in seinen Beziehungen zur Entwicklungsmechanik," *Archiv für Entw.-Mechanik,* Vol. 24, pp. 677–83. 1907.

———. "Noch einige Bemerkungen zur Chtonoblastenfrage," *Archiv für Entw.-Mechanik,* Vol. 26, pp. 178–87. 1908.

Naegeli, C. W. *Mechanisch-physiologische Theorie der Abstammungslehre.* Munich, 1884.

Nordenskiöld, Erik. *Biologiens historia,* Vol. III. Helsingfors, 1924.

Otto, R. and Munter, H. "Bakteriophagie," *Hdb. d. path. Mikroorg.,* Vol. 1, No. 1.

Pasteur, Louis. "Fermentations et générations dites spontanées," *Oeuvres,* Vol. 2. Paris, 1922.

———. "Etudes sur le vinaigre et sur le vin," *Oeuvres,* Vol. 3. Paris, 1924.

———. "Etudes sur la bière," *Oeuvres,* Vol. 5. Paris, 1928.

———. "Maladies virulentes, virus—vaccins et prophylaxie de la rage," *Oeuvres,* Vol. 6. Paris, 1933.

———. "Die in der Atmosphäre vorhandenen organisierten Körperchen, Prüfung der Lehre von der Urzeugung," translated by Dr. A. Wieler, in Ostwald's *Klassiker der exakten Wissenschaften,* No. 39. Leipzig, 1892.

Pflueger, Eduard. "Über die physiologische Verbrennung in den lebendigen Organismen," *Pflueger's Archiv,* Vol. X, p. 251. 1875.

Preyer, Wilhelm. *Naturwissenschaftliche Tatsachen und Probleme.* Berlin, 1880.

Przibram, Hans. "Kristallanalogien zur Entwicklungsmechanik der Organismen," *Archiv für Entw.-Mechanik,* Vol. 22, pp. 207 ff. 1906.

———. "Vitalität," *Experimentalzoologie,* Vol. 4. Leipzig/Vienna, 1913.

———. *Temperatur und Temperatoren im Tierreiche.* 1923.

———. *Die anorganischen Grenzgebiete der Biologie, insbesondere der Kristallvergleich.* Berlin, 1926.

Putzeys, Felix. "Über die Abiogenesis Huizingas," *Pflueger's Archiv,* Vol. IX, p. 391 (1874), and *Pflueger's Archiv,* Vol. XI, p. 387 (1875).

Rhumbler, Ludwig. "Aus dem Lückengebiet zwischen organismischer und anorganismischer Materie," *Ergebnisse der Anatomie und Entwicklungsgeschichte*, Vol. 15, pp. 1–38. Wiesbaden, 1906.

———. "Das Protoplasma als physikalisches System," *Ergebnisse der Physiologie*, Vol. 14. 1914.

———. "Methodik der Nachahmung von Lebensvorgangen durch physikalische Konstellationen," *Handbuch der biologischen Arbeitsmethoden* (E. Abderhalden), section 5, part 3A. Berlin, 1923.

———. "Anorganisch-organismische Grenzfragen des Lebens," in Driesch-Woltereck's *Das Lebensproblem*. 1931.

Rinne, Friedrich. *Grenzfragen des Lebens*. Leipzig, 1931.

———. "Beiträge zur biologischen Kristallographie," *Cbl. f. Min.* 1931.
  I. Diskussion eines Referates über "Spermien als lebende flüssige Kristalle," pp. 233 ff.
  II. Vergleichende Vermerke über morphologisch-physiologische Gliederungen im Bau des Organischen und Anorganischen, pp. 273 ff.
  III. Inhomogenität und Pseudohomogenität bei organischer und anorganischer Materie, pp. 305 ff.
  IV. Zur Nomenklatur der Hauptstufen des Feinbaus, pp. 337 ff.

Roux, Wilhelm. "Der züchtende Kampf der Teile oder die 'Teilauslese' im Organismus, Zugleich eine Theorie der 'funktionellen Anpassung.'" Reprinted in *Gesammelte Abhandlungen über Entwicklungsmechanik der Organismen*, Vol. I, p. 409 ff. Leipzig, 1895.

———. "Die Entwicklungsmechanik, ein neuer Zweig der biologischen Wissenschaft," *Vorträge und Aufsätze über Entwicklungsmechanik der Organismen*, No. 1. Leipzig, 1905.

Samuelson, Paul. "Über Abiogenesis," *Pflueger's Archiv*, Vol. 8, pp. 277 ff. 1874.

Schmidt, W. I. "Gewebe der Tiere, submikroskopischer Bau," *HWB d. Nat.-Wiss.*, Vol. V, pp. 167–79. Jena, 1935.

Schmucker, T. *Geschichte der Biologie*. Göttingen, 1936.

Teichmann, E. and Rhumbler, L. "Urzeugung," *HWB d. Nat.-Wiss.*, Vol. X, pp. 110 ff. Jena, 1935.

Tschermak, A. v. *Allgemeine Physiologie*, Vol. I, Part 1, No. 2. Berlin, 1916/1924.

Tyndall, John (and Pasteur, L.). *Les microbes organisés, leur rôle dans la fermentation, la putréfaction et la contagion*. Paris, 1878; Johnson Repro, 1966.

Uexkuell, J. v. "Definition des Lebens und des Organismus," *Handbuch der normalen und pathologischen Physiologie*, Vol. I. Berlin, 1927.

———. *Theoretische Biologie*, 2nd ed. Berlin, 1928.

Verworn, Max. *Allgemeine Physiologie*, 5th ed. Jena 1909.

Weiss, B. "Zum Urzeugungsproblem," *Zentralblatt für Physiologie*, Vol. XXI, p. 74. 1907.

### ADDENDUM

Kanitz, Aristides. "Obere Temperaturgrenze des Lebens," *Tabulae biologicae*, edited by Oppenheimer and Pincussen, Vol. II. Berlin, 1925.

---

The works of Wilhelm Reich are published in cooperation with The Wilhelm Reich Infant Trust Fund. Those seeking additional information are advised to contact the trust fund at 382 Burns Street, Forest Hills, N.Y. 11375, or The Wilhelm Reich Museum, Orgonon, Rangeley, Maine 04970.

CPSIA information can be obtained at www.ICGtesting.com
Printed in the USA
LVOW05s0038260713

344740LV00004B/117/A